Mark Beckner
with Scott McFarland
Administering, Configuring, and Maintaining Microsoft Dynamics 365 in the Cloud

Mark Beckner
with Scott McFarland

Administering, Configuring, and Maintaining Microsoft Dynamics 365 in the Cloud

DE
G
PRESS

ISBN 978-1-5474-1611-0
e-ISBN (PDF) 978-1-5474-0006-5
e-ISBN (EPUB) 978-1-5474-0008-9

Library of Congress Cataloging-in-Publication Data
A CIP catalog record for this book has been applied for at the Library of Congress.

Bibliographic information published by the Deutsche Nationalbibliothek
The Deutsche Nationalbibliothek lists this publication in the Deutsche Nationalbibliografie; detailed bibliographic data are available on the Internet at http://dnb.dnb.de.

Published by Walter de Gruyter Inc., Boston/Berlin
Printing and binding: CPI books GmbH, Leck
♾ Printed on acid-free paper
Printed in Germany

www.degruyter.com

Acknowledgments

We would like to thank those who helped us make sure this book was as good as it could be. We would like to thank Jennifer Curiak for her fine work serving as the tech reviewer and making sure that every step in the book was correct. Mary Sudul did a great job as copyeditor of refining every word. We would also like to thank Jeffrey Pepper and Jaya Dalal for their edits and helping us through the process and the excellent work of Angie MacAllister and her team in producing a clean book.

About the Authors

Mark Beckner has been working with companies for nearly 20 years building out integrations, custom solutions, portals, and CRM/Dynamics 365 implementations. He has worked with the Dynamics platform since it first came out as Microsoft CRM 1.0, and has been involved in architecture, development, and support on it for many customers. He works with a variety of platforms that integrate with CRM, including BizTalk, Scribe, SSIS, SSRS, and .NET applications. He is a developer, architect, and technical strategist, and works to ensure that his clients are implementing solutions and platforms in an efficient and productive manner. He lives with his wife, Sara, and their two boys Ciro and Iyer Blue on a small farm in Western Colorado. You can learn more at www.inotekgroup.com, and can always reach out to him directly at mbeckner@inotekgroup.com.

Scott McFarland has had a wide and varied career. From his early days managing the kitchen operations of McFarland's Restaurant to his experiences as an independent business and technology consultant, he delivers outstanding results to his clients. Educationally, Scott achieved a Bachelor's of Science in Physics, graduating Cum Laude and later earning his Masters of Business Administration in Finance & Operations Management. Using his educational background plus his hands-on experience in technology, he partners with clients to vastly improve their business operations through managing people changes, implementing efficient processes and matching technology to the needs of the business. During the last few years, Mr. McFarland has earned many certifications including Microsoft Dynamics CRM, ITIL Foundation, Project Management Professional, SQL Administration, Microsoft Business Professional, EDI and many other Microsoft certifications.

Contents

Introduction

With just 8 hours of your time, using the pages in this book, you can achieve success with Dynamics 365.

It seems that virtually every organization is adopting Dynamics 365. Large or small, companies are realizing that they need a coherent strategy to manage their customer relations, and they need a platform on which they can model their business and support their business processes. Management within these organizations understand that Dynamics is the platform that will allow them to revolutionize how they interact with their customers and across teams within the company. However, understanding that the platform is needed, and properly implementing it are two very different things.

You are picking this book up because you either have Dynamics or are considering acquiring it soon. You have made a correct choice in technology – it really can do everything you need it to do. It does have the capacity to support you in whatever direction you decide to take. But, it is your responsibility to ensure that it is implemented properly. You must become familiar with this product if you want to be successful with this product!

Our goal in producing this book is to provide you with an overview of Dynamics that will allow you to quickly work through most aspects of the platform, whether you are a developer or not. It is imperative that you know how the platform functions, and understand what features and components are available to you. You should never move forward with implementing a product that is a black box to you. If you are an executive that is directing your IT team to implement Dynamics, or you are a director that is working with external consultants to build out your system, you must do everyone a service and become familiar with the platform.

You can take the following steps to achieve success and become acquainted with Dynamics 365:

1. Set aside 8 hours this week.
2. Work through the steps in each chapter of this book, and do the hands on clicking that each chapter outlines.
3. Even if you have never developed anything in your life, walk through the basics of creating components and downloading the tools as outlined in these pages.
4. Walk away with a solid understanding of this platform in order to deploy or expand your Dynamics 365 environment.

We've worked with many different organizations to build out their CRM environments. Sometimes these are small projects, and sometimes these are projects that consist of many resources and cost millions of dollars. Regardless of the end price, you have an obligation to yourself and to those who are trusting you to build out this Dynamics solution. With just a few hours of your time, you can understand this platform.

Our hope is that you use this book, and that by having this framework you save yourself a tremendous amount of stress – there is nothing more stressful than managing a team building something that you don't understand. If you understand Dynamics, you can direct your team to implement it. You can save your organization time and very substantial amounts of money.

You have the solution in your hand now – set aside a day this week and get your hands dirty! Work with Dynamics, understand what a great and powerful tool it is, and be successful with your solution!

And, of course, if you have any questions, you can always reach out to us.

Mark Beckner—mbeckner@inotekgroup.com 970.640.5419

Scott McFarland—scott.mcfarland@controltowerconsulting.com 503.989.0232

Chapter 1
Set Up and Configuration

This chapter outlines the steps necessary to set up and access a new Microsoft Dynamics 365 Online instance. It also covers basic navigation as well as system configuration and auditing.

Setting Up Dynamics 365 Online

Begin by navigating to the following URL:

https://www.microsoft.com/en-us/dynamics/free-crm-trial.aspx

For years, there has a been a 30-day trial offered as part of the sign up. This trial version has all of the functionality of the paid version, and allows you time to explore the product and set up the basic functionality and user base before you have to pay for the product. Once the 30-day trial expires, you will be re-directed to a page where you can sign up and enter in billing details—no data or configurations that you have done during the trial period will be lost. When the page shown in Figure 1.1 opens, click on the Get Started button.

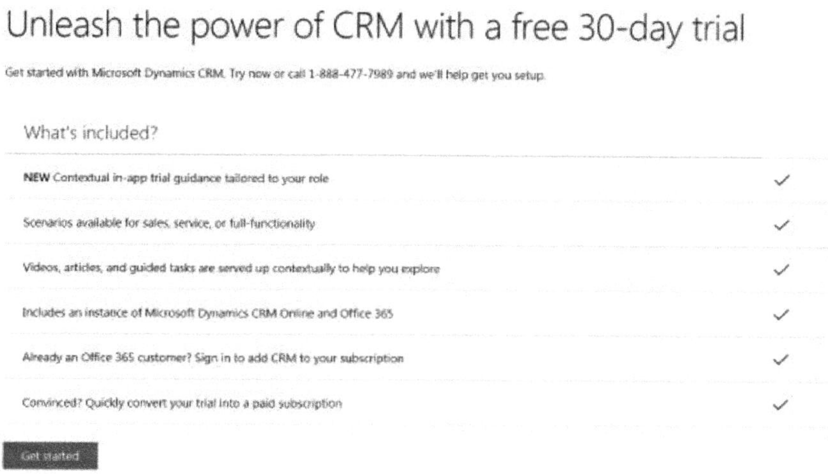

Figure 1.1: Accessing a free trial of Dynamics 365.

DOI 10.1515/9781547400065-001

After clicking the “Get started” button, you will be asked to select the type of configuration you would like for CRM. However, you can skip the marketing and click on the “Sign up here” link hidden on the page, as shown in Figure 1.2. Scroll down to the bottom of the page, the link is below the tiles and above the “Work email” entry box.

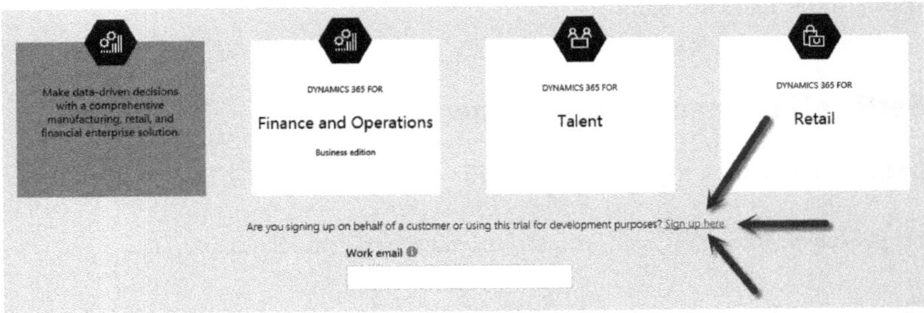

Figure 1.2: The sign up link is pretty well hidden on the screen—don’t miss it!

On the next page (titled “Welcome, let’s get to know you”), enter in the details of your company and click Next (see Figure 1.3). Take care to enter your information from this point on accurately, as there is no “Back” button that will allow you to revise your information.

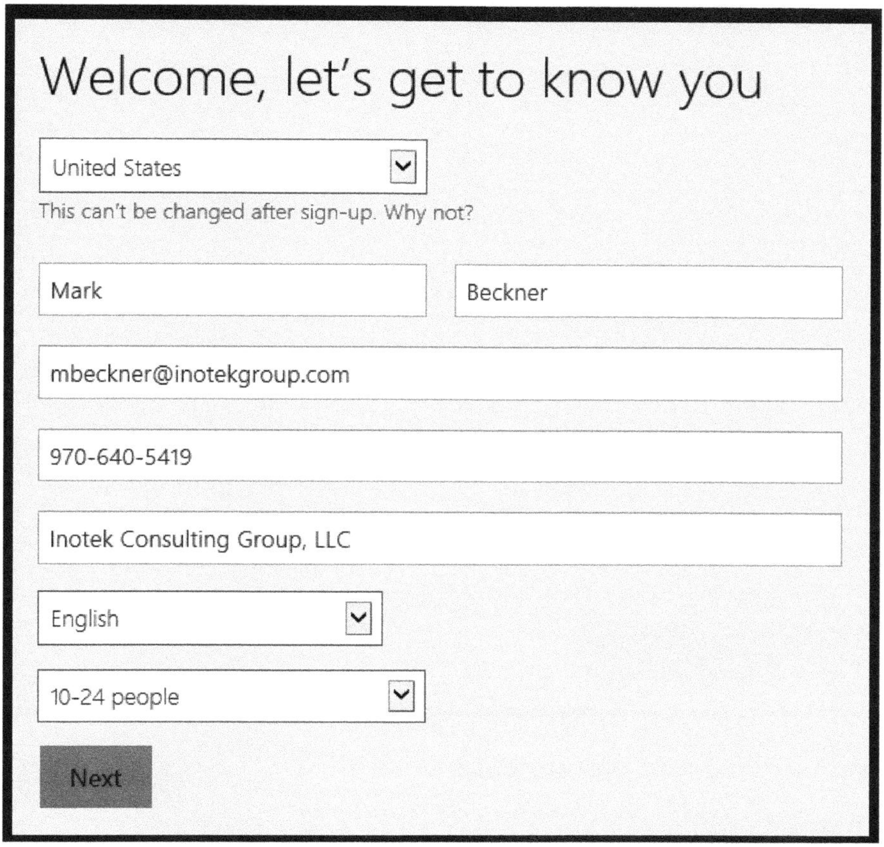

Figure 1.3: Basic setup information.

The next screen (Figure 1.4) will allow you to create your user ID and password. The domain name you choose will be visible to anyone who is navigating to your CRM instance. Make sure that you use something that you want to have people see when telling them how to navigate to your site. Additionally, all users you create will have the same "*@domain*.onmicrosoft.com" in their login—choose your name wisely here, since it will be with you long-term once you create your Dynamics instance.

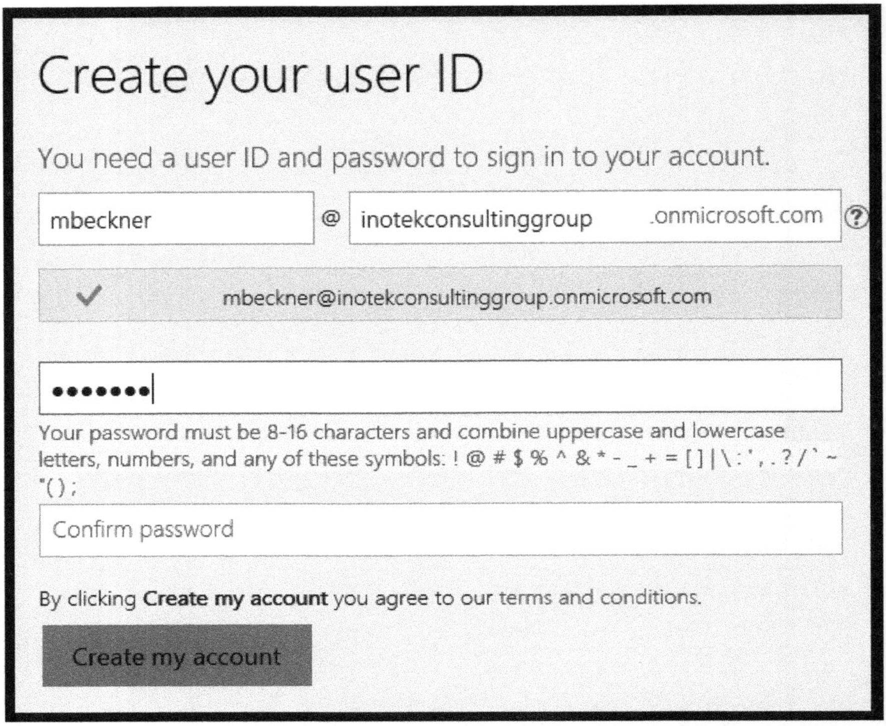

Figure 1.4: Entering in your account credentials.

Click through the remaining pages, and complete the confirmation process. Once you are at the end of these pages, your D365 instance will be created and you'll be able to log in and select the type of CRM instance you want to start with. There are certain modules that are available for different types of organizations, and Microsoft allows you to select a basic template and module configuration based on your type of business (see Figure 1.5). The safest thing to do is to select the "All of these" option, which will ensure you have easy access to all of the functionality available in Dynamics 365. Click the "Complete Setup" button at the bottom of the screen when you are ready to proceed.

Figure 1.5: Determine which installation is most appropriate for you.

After you select which setup scenario you want, some additional configuration will occur and eventually you will be directed to the main Dashboard page of CRM (see Figure 1.6). At this point, your sign up process is complete and you have access to a fully functional D365 instance.

You can convert your free trial to a paid subscription at any time. The easiest way to do this is via the "Buy Now" button at the top of the screen. Before paying, you should use up your 30 free days!

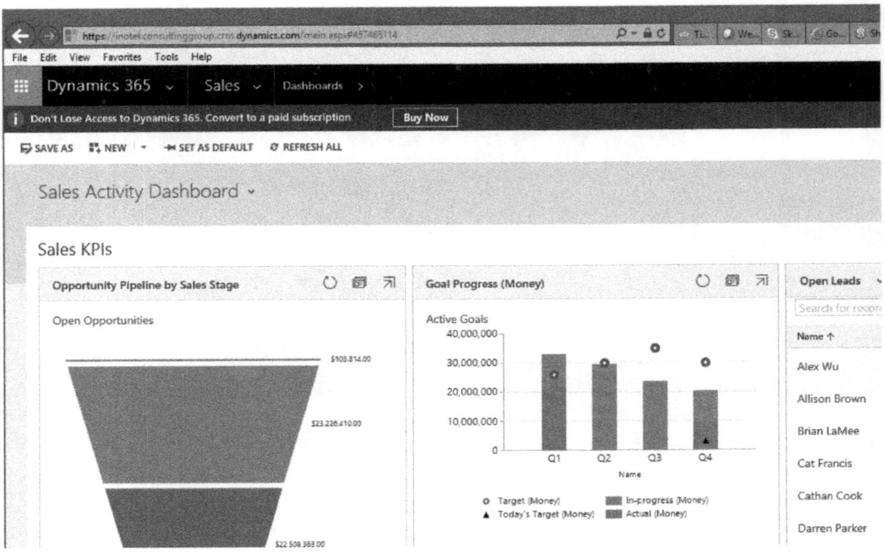

Figure 1.6: The main landing page in Dynamics 365—the Sales Activity Dashboard.

To log into the site in the future, browse directly to your D365 account by typing in the URL that is formatted like this: https://*inotekconsultinggroup*.crm.dynamics.com. If you browse to the root portal.office.com and enter your credentials, you will likely end up on a generic administrative landing page (shown in Figure 1.7). Signing up for Dynamics 365 gives you access to all of the functionality of Office 365, including downloadable versions of Word, Excel, and other Office products. If you click on Dynamics 365 to get to your D365 instance from here, you likely won't be able to access the main CRM landing page that you would expect.

Figure 1.7: Signing up for D365 also gives you access to all of the Office 365 products.

Basic Navigation of D365

Navigating CRM successfully and efficiently can require some time to learn. There are so many ways to get around the system, and so many options that many users become frustrated. This section gives a simple overview of the basics of traditional navigation, but many users may want to skip the navigation options and use the Advanced Find tool (covered in the next section of this chapter.)

Your ribbon menu may look different than what is laid out in the images in this section. The menu navigation options will depend on what features have been enabled and the exact version of Dynamics that you are using. Figure 1.8 shows one potential configuration of the menu. Figure 1.9 shows a second view. In the first case, menu options will be under the multiple lines next to the title. In the second option, clicking on the drop-down next to the second option (in this image, that is Sales, but this is a dynamic value) will present you with the full menu.

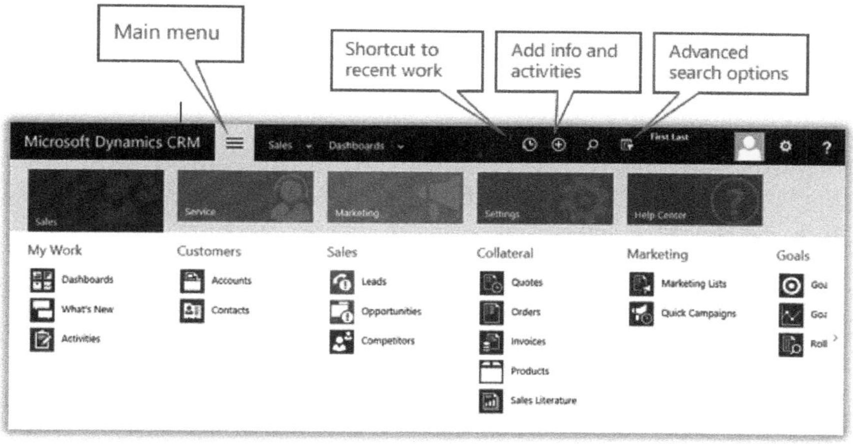

Figure 1.8: One view of the navigation menu.

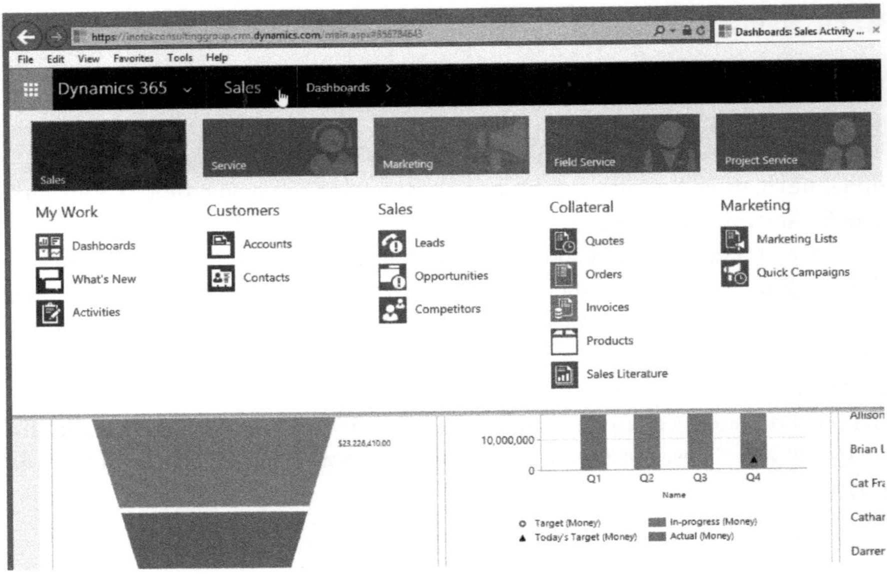

Figure 1.9: Another view of the navigation menu—layout is often changing, and may look different on your instance.

From this main toolbar, you will be able to access virtually all of the entities you'll be working with. Finding out where they are located can take a little work, but as soon as you spot what you are after, you can click on it. Figure 1.10 shows where Accounts is located in the toolbar.

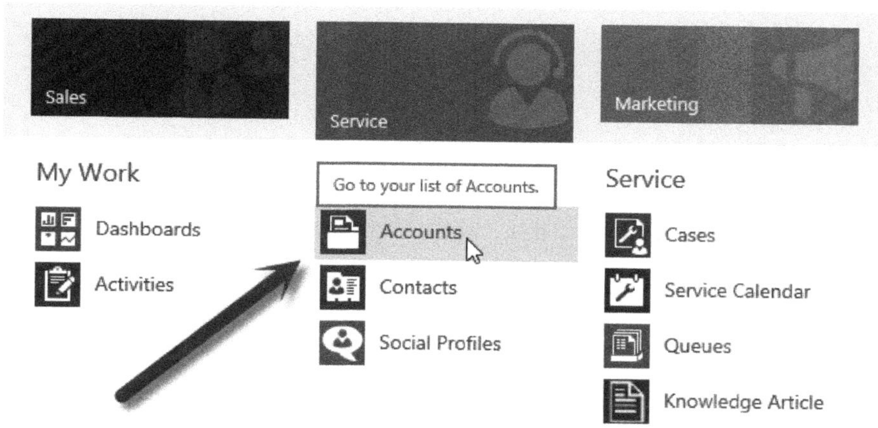

Figure 1.10: Accessing Accounts.

Once you've clicked on an entity, you will be presented with a view of data. There are many views available to you. By default, you will most likely see "My Active Accounts" when you open the account record. You may or may not have data that shows in this view. This will depend on whether you are listed as the Owner on any account records. You can change the view of data by clicking the drop-down next to "My Active Accounts," which will show all of the System Views (such as All Accounts, which will show everything regardless of ownership) and will also allow you to create a new Personal View. Figure 1.11 shows all the standard views that are available on the account form.

Figure 1.11: The many views available to you from the Account entity.

Clicking Create Personal View allows you to set up your own way of visualizing data based on filters that you determine. Most of the core entities, like Accounts, Opportunities, and Leads have extensive out-of-the-box views available to you, but custom entities will have generic views that may require you to create your own view into the data. Clicking Create Personal View will open up an Advanced Find window, where you can set whatever filter criteria you would like and save it—once saved, you'll be able to access this view via the drop-down any time in the future.

Advanced Find

One of the easiest ways to quickly navigate around CRM and find the exact record or records you are looking for is to use the Advanced Find utility. This lets you select the specific entity you are looking for along with the filter criteria for finding the exact record you are after. Advanced Find is somewhat hidden in the toolbar, but appears as the funnel shape in the upper right section. Clicking this funnel icon will open an Advanced Find window, as shown in Figure 1.12.

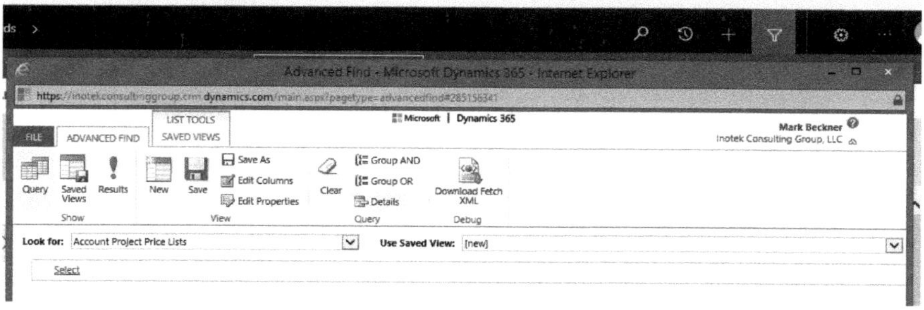

Figure 1.12: The Advanced Find window—become familiar with this powerful tool!

In order to find records, you can follow this example: If you are trying to find all of the Accounts that are available, you select Accounts in the "Look for property" field, and then click the Results button (the red exclamation point) in the toolbar. This would bring back all Accounts, as shown in Figure 1.13.

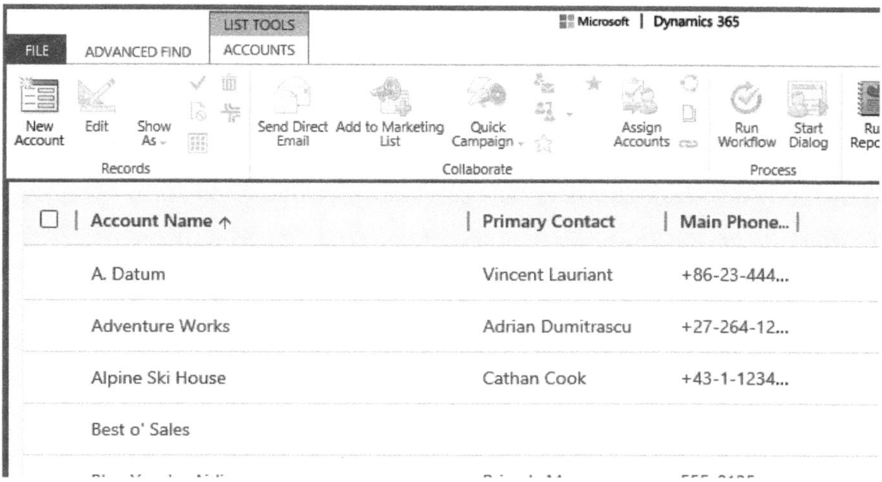

Figure 1.13: The result set window within the Advanced Find tool.

However, you most likely will want to filter your results. To do this, click back on the Advanced Find tab and click the "Select" option underneath Accounts and select a field that you want to filter on. You can select as many fields as you want to create the filter you are after. Figure 1.14 shows a simple query created to limit the result set to only those accounts that have a name that starts with "M," and that were created on or before a specific date.

There are two create dates—Record Created On and Created On. The Record Created On always records the actual system date on which a record was created. The Created On date will record this date if the record is manually created, but if it has been created through code (such as a plugin) or an import, this date may reflect an overridden value (for example, during import you may want to capture the original create date from an old system).

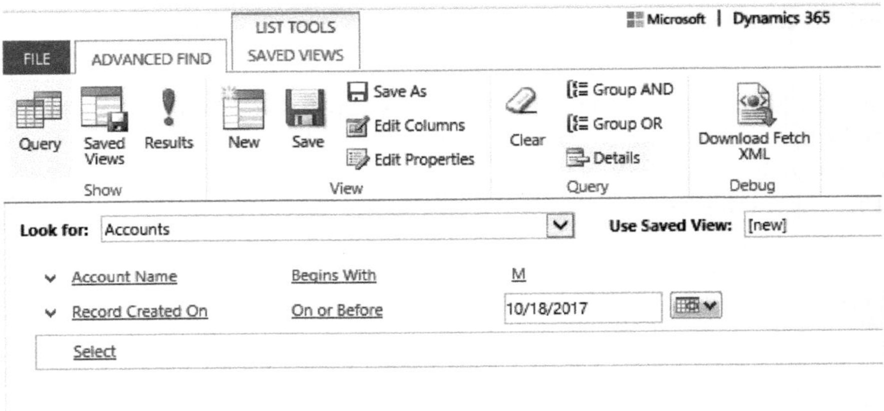

Figure 1.14: Adding filters within Advanced Find.

Once you have your results, you can either open the record you are looking for, or you can export your records. On the results page, click on the "Export Accounts" tab (this tab name will change depending on the entity you are searching), which should be the last button on the right of your toolbar. This will give you several options (see Figure 1.15 for reference).

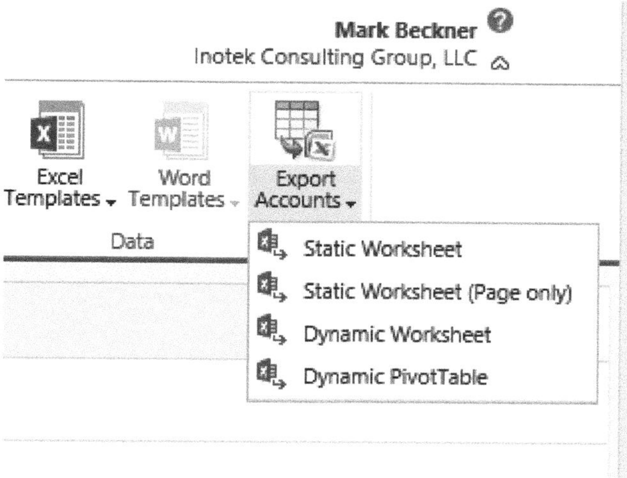

Figure 1.15: You have several options when exporting your result sets.

1. *Static Worksheet*—This will export all records that are in the list of results. If you have multiple pages of results, then this will export ALL your records to a static Excel sheet. Note that the export is limited to 100,000 records. If you have more records in your Advanced Find than that, you will need to figure out how to limit your result sets to smaller than 100K each and do multiple exports.
2. *Static Worksheet (Page only)*—This will export exactly what is shown in the result set on your screen.
3. *Dynamic Worksheet*—You can export your current result set and keep it connected to Dynamics 365 via Excel after the export. For example, let's say that you have 10 account records in CRM, and you export a dynamic spreadsheet. The spreadsheet with show the 10 account records. Now, let's say you go into CRM and add another account. When you open your spreadsheet again, it will get the updated information and will now show 11 accounts.
4. *Dynamic PivotTable*—You can select this option for advanced Excel work, which will allow you to stay connected to CRM as a data source and use Excel to manipulate the result sets in many different ways.

If you need to revise the fields that are returned in your record sets, you can change the fields by clicking the "Edit Columns" button on the ribbon on the main Advanced Find tab. Clicking this will open a small window where you can add/remove fields as well as alter the sorting order of your results. In Figure 1.16, the Account Number field list is being added by selecting it and clicking OK. Once the field has been added, it can be moved around by selecting it and clicking the arrows. Sorting can be altered by clicking on the Configure Sorting tab.

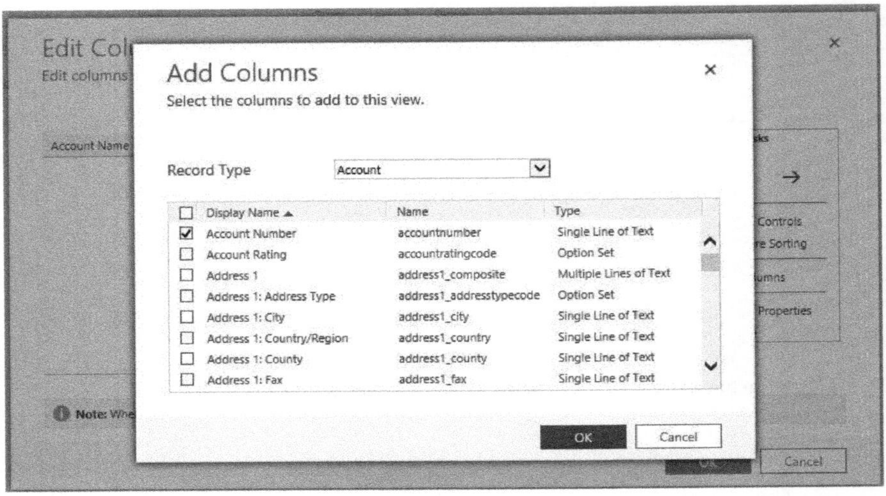

Figure 1.16: You can revise the columns in your view.

The Advanced Find tool is very powerful and can greatly simplify finding records in CRM.

System Settings and Auditing

While you have a fully functional system ready to go out of the box, there are several core configurations that you'll want to focus on before proceeding. All configurations are available via the Settings menu. Figure 1.17 shows the various options available from Settings. Here you can make customizations to CRM, add users and configure security, set up workflows, import data, and do a variety of other administrative activities.

Figure 1.17: The Settings tab.

We will look at security in the next chapter (Users, Roles, Teams, and the like), so for now we'll look at the system administration settings. This section (under the System heading on the Administration tab) is shown in Figure 1.18. To access this page, open the Settings tab, and under the System heading, click Administration. The various options here can be used to change general system-level settings.

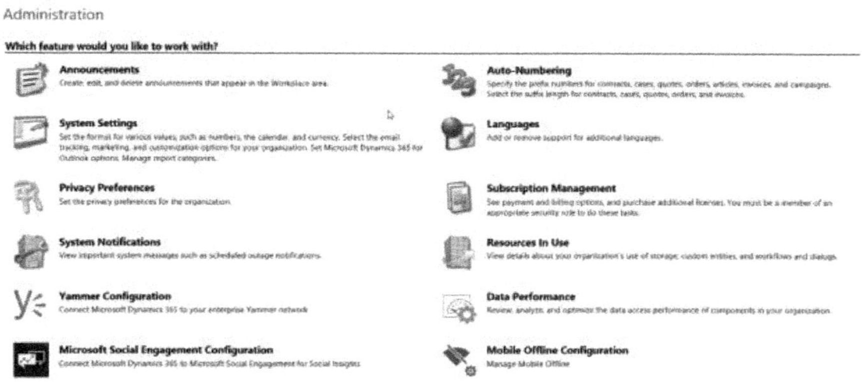

Figure 1.18: A number of options are available to you from the administration screen.

Clicking on System Settings will give you access to setting a wide variety of global properties. As you can see in Figure 1.19, there are many tabs and properties that can be set.

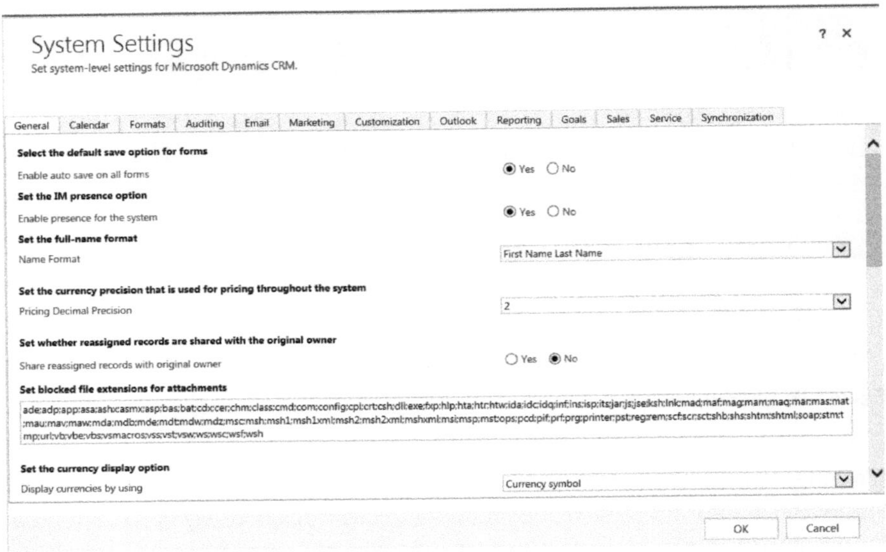

Figure 1.19: There are many tabs and settings available within System Settings.

The property regarding the default save options for forms is a common one to set. Many companies do not want forms to automatically save. In order to prevent automatic saving, click on the No option for the "Enable auto save on all forms" property. Once set, this will require that users click the Save button that is almost hidden in the lower right hand corner of your form (see Figure 1.20).

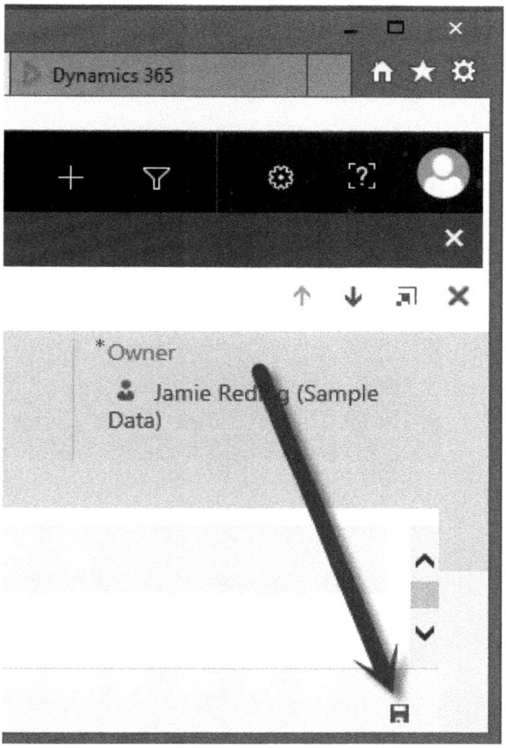

Figure 1.20: If you disable auto save, users will have to click the tiny save button at the bottom of a form to save their data.

Auditing is an important global setting as well. The ability to track the history of records is invaluable. With CRM auditing, you can track what fields were changed on what entity by what user over time. Historically, many organizations would not enable auditing due to concerns about the amount of data that would be stored to the database, but with a Dynamics 365 Online instance, where everything is stored in the cloud, concerns about disk space are irrelevant. Troubleshooting issues on forms with complex JavaScript, business logic, or plugins is made much simpler by being able to see what activities occurred, and whether they were made by human users or system users (code and workflows).

In order to enable general auditing, click on the Auditing tab in System Settings. In Figure 1.21, you can enable auditing functionality by selecting Start Auditing, as well as selecting which areas and entities to enable auditing within. Make sure to click OK in order to save the global settings.

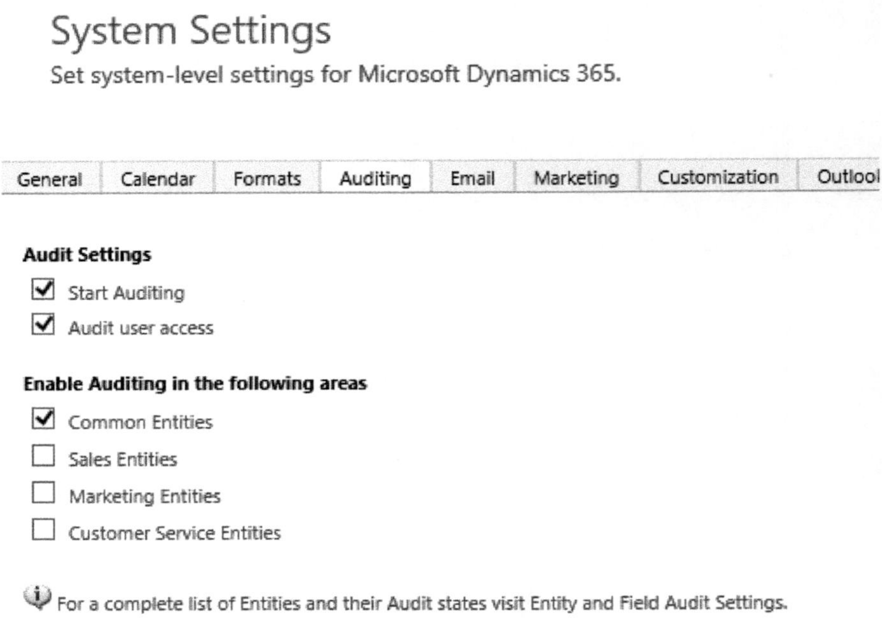

Figure 1.21: Enabling system-wide auditing.

Once the global settings have been set, you can click on Settings in the main toolbar (see Figure 1.22).

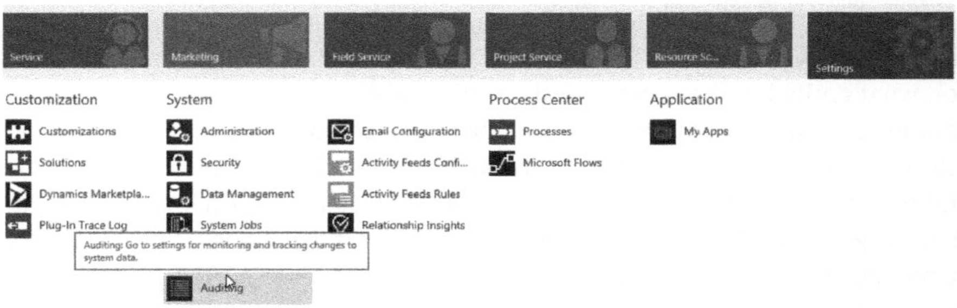

Figure 1.22: Accessing auditing from the main ribbon.

From the page that opens (shown in Figure 1.23), you can quickly navigate to all of the auditing functionality available to you.

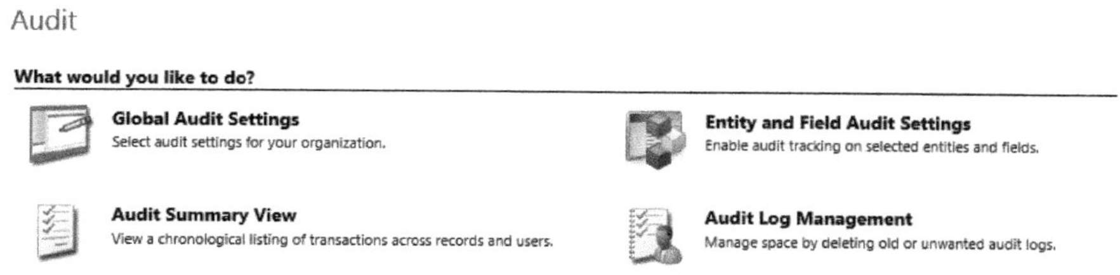

Figure 1.23: Audit functionality.

Clicking on Global Audit Settings will take you back to the tab previously noted. Clicking Entity and Field Audit Settings will open the customizations window where you can click on individual entities and fields of those entities to set whether these will be audited. From the Solution Default Solution list, select the applicable entity to display the configuration pane for that entity. (Figure 1.24 shows where this setting is within the configuration pane).

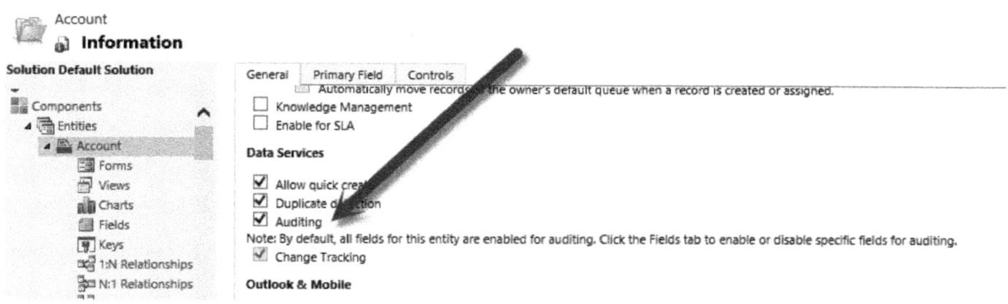

Figure 1.24: You can enable or disable auditing on individual entities.

To audit an individual field, click on the Fields tab underneath the entity, which will display all of the fields on that entity. Next select the individual field you are interested in and click Edit. Yet another window will open. In this newly opened window, you will be able to set whether auditing will be performed on this specific field or not (see Figure 1.25). By default, fields will be set to have auditing enabled, and new fields that you create will default to auditing enabled as well.

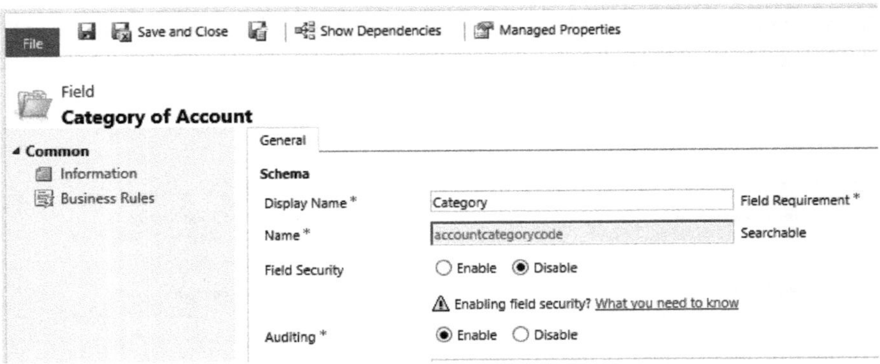

Figure 1.25: You can also set auditing on individual fields.

With auditing enabled, you can view activities on your system by clicking on the Audit Summary View option (refer back to Figure 1.23 above for where this is located). The Audit Summary will display every activity that is being tracked in the system. Figure 1.26 shows all the actions that have taken place. Clicking on these individual actions will provide more detail as to what actually occurred.

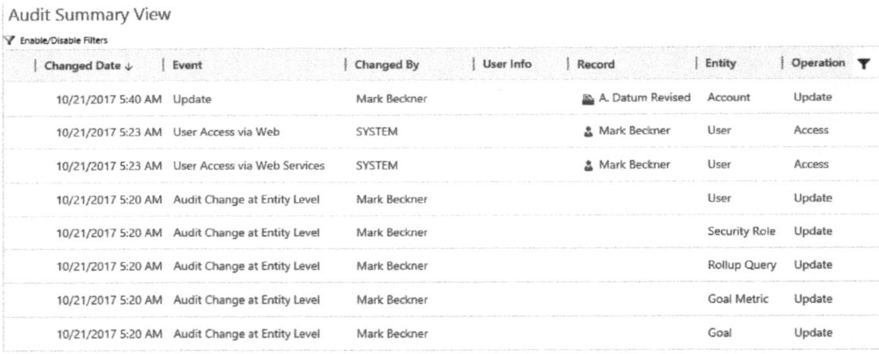

Figure 1.26: A summary of all system activities.

You can view audit history directly from a specific record as well. For example, if you have a specific account record open, and you want to see what kind of activity has taken place on that record, you could click the drop-down next to the account name and select Audit History, as shown in Figure 1.27.

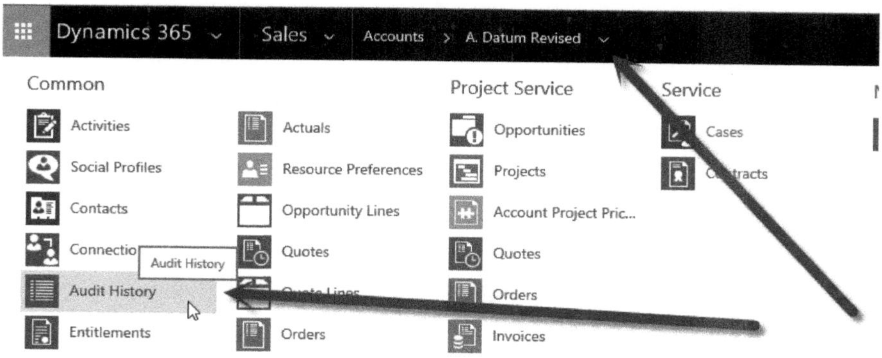

Figure 1.27: Accessing the history of activity on a specific record.

Clicking Audit History on the record will show all the activity that has taken place on that entity from the time that auditing was enabled. Figure 1.28 shows that the Account Name and the Main Phone fields were changed in this example. It shows what the original value was and what the new update value is, along with what user made the change and what time it was changed.

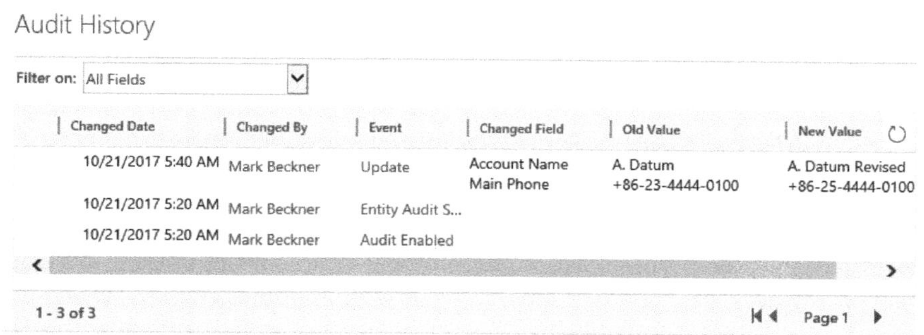

Figure 1.28: The audit shows changes to specific fields, along with a timestamp.

Auditing can be a fantastic aid in troubleshooting issues and determining what took place on a record over time. Every organization should enable auditing as one of the very first steps of setting up their Dynamics environment.

Auto-Numbering

For some of the core entities, Dynamics has the ability to create a number using predefined formatting. Key information about out-of-the-box auto-numbering includes:

1. Creates auto numbers with prefixes and zero padding.
2. Works with the core default entities of Contracts, Cases, Articles, Quotes, Orders, Invoices, Campaigns, Categories, and Knowledge Articles.
3. Is easy to maintain and configure. Simply click on the Auto-Numbering option within Settings on the Administration tab and set values as shown in Figure 1.29.

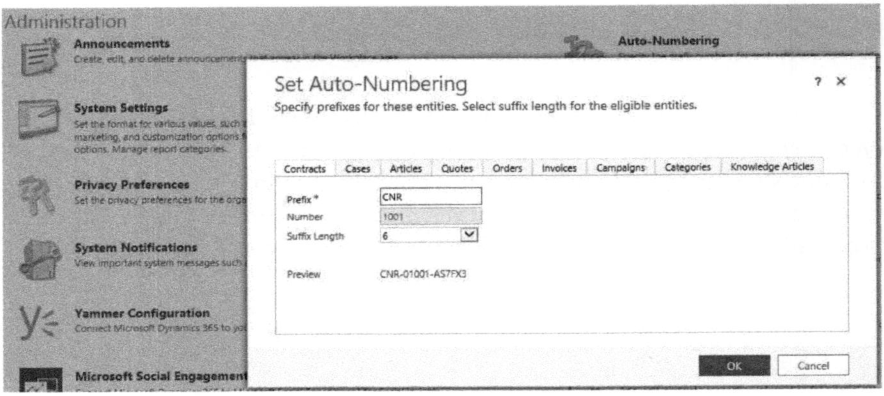

Figure 1.29: Out-of-the-box auto-numbering is available from the Administration tab in Settings.

There will be many scenarios where you will want to control your auto-numbering format in more detail, or apply auto-numbering to custom entities that you create. In these cases, you will have to work with a plugin solution. There are several free options available for download via CodePlex and other online sources, or you can write your own.

Summary

This chapter provided a basic overview to installation and navigation of Dynamics 365, along with an introduction to configuring auditing across entities and fields. In the next chapter, you will look at user settings and security across the system.

Chapter 2
Security

This chapter outlines the use of Business Units, Security Roles, Teams, Users, and Field Level Security, and other topics related to security within your Dynamics 365 Online environment. In order to access the various areas for configuration of security, click on Settings from the main toolbar, and under the System group, select Security. The screen shown in Figure 2.1 will be displayed. From here you can access everything related to security.

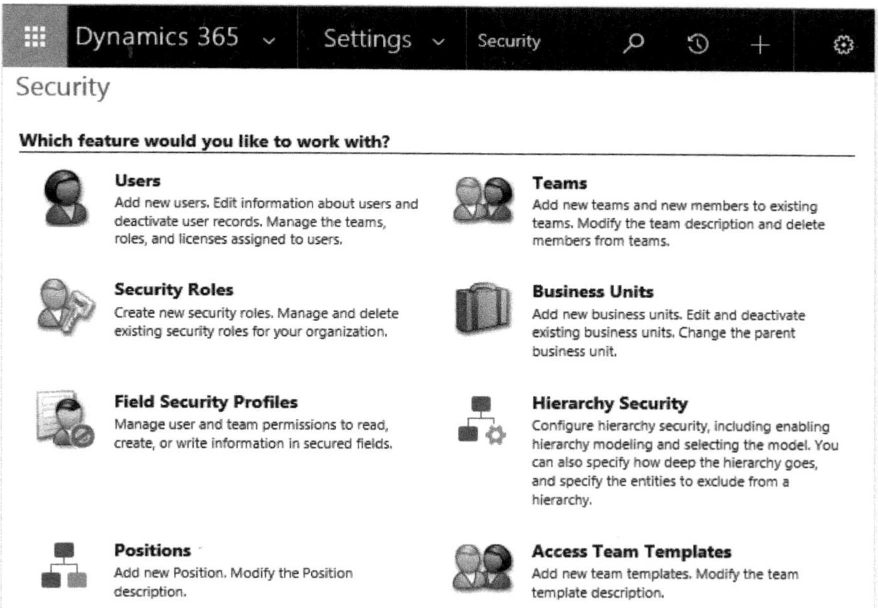

Figure 2.1: The main security screen with various options.

Business Units

Business units are the foundation of the security structure in Microsoft Dynamics 365 Each user in the CRM must be part of a business unit. There is a default business unit that is created when D365 is installed. This is called the root business unit, and it cannot be deleted or disabled—only renamed. More business units

DOI 10.1515/9781547400065-002

can be created if different levels of access to information are required for different groups within your organization.

There must be at least one business unit in D365 and only one root business unit. The root business unit is the topmost point of the CRM organizational hierarchy and all other business units are its children. The root business unit is usually the corporation or the holding company and the child business units are subsidiaries, divisions, or departments of the business.

As you can see in Figure 2.2, creating a new business unit requires that there be a parent business unit defined. In this image, the window in the background shows the default business unit created automatically during the initial setup of CRM. In the window in the foreground, a new business unit is being created. The Parent Business field has been automatically populated. You can switch this to another business unit (if you have one to select from), but you cannot have two top-level business units.

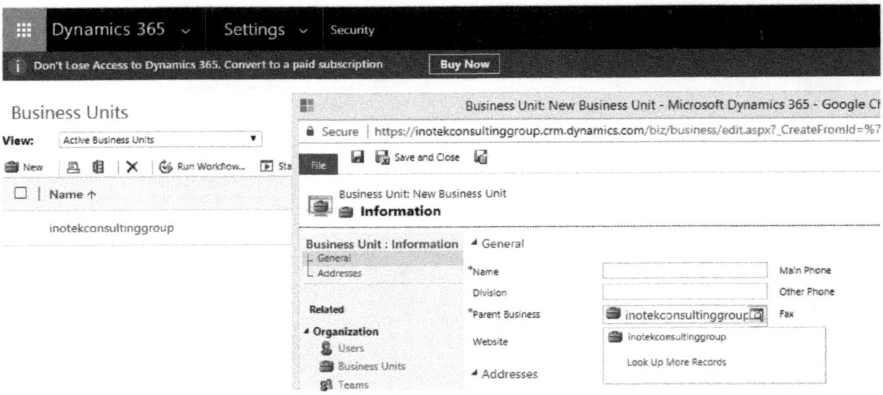

Figure 2.2: Creating a new business unit.

Most companies will only have a single business unit. If you have advanced security needs, have a CRM that models data for multiple companies, or have a complex organization, you will want to use a hierarchy, otherwise a single business unit should suffice.

Security Roles

A *security role* defines how different users (such as salespeople, administrators, customer service representatives, and developers of CRM) can access different types of records. To control access to data, you can modify existing security roles, create new security roles, or change which security roles are assigned to each user. Each user can have multiple security roles.

Security role privileges are cumulative: having more than one security role gives a user every privilege available in each of those roles. Each security role consists of record-level privileges and task-based privileges. Record-level privileges define which tasks a user with access to the record can do, and include those shown in the following list.

Each of the following privileges are available based on the access level defined in the security role. For example, you can restrict users to be able to *read* records across all entities, but only *create* records within a certain entity.

- Read—Users can read records.
- Create—Users can create records.
- Delete—Users can delete records.
- Write—Users can update records.
- Assign—Users can assign records to other users (causing those users to become the *owners* of the records).
- Share—Users can share their information with others.
- Append—Users can attach information to an existing record, such as an activity or note.
- Append To—Users can attach the record they are on to an existing record (for example, when creating a new note, they can attach it to an existing opportunity).

CRM comes with default security roles already configured. These roles can be assigned as is or copied and modified to address a company's specific security needs. To work with security roles, go to the main Settings tab, and in the System group, select Security, then select Security Roles. You can create a new role by either clicking on the "New" button, or by selecting an existing role and click More Actions and then Copy Role (see Figure 2.3).

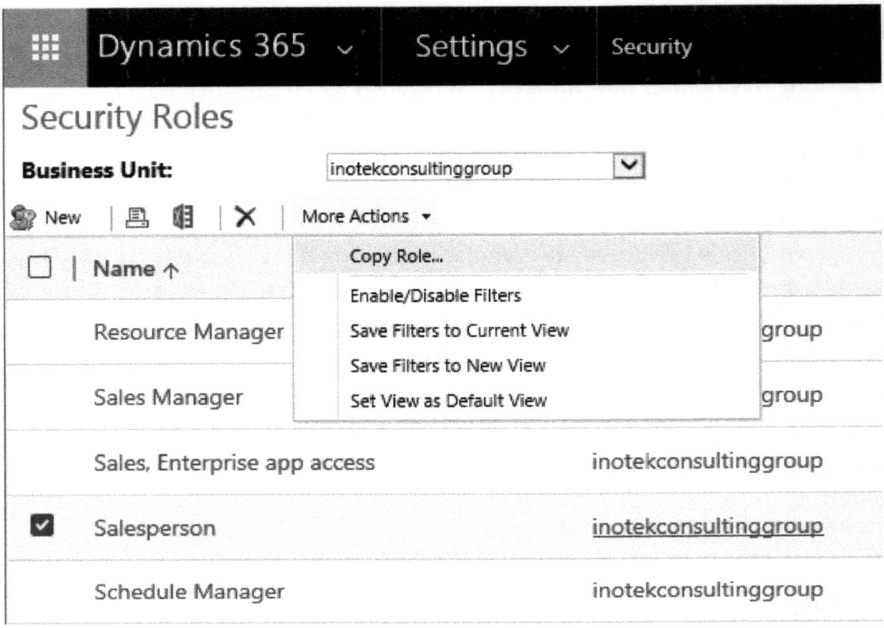

Figure 2.3: You can copy a role as a starting point rather than configuring a new role entirely from scratch.

Copying a role allows you to leverage all of the security settings that have been configured on the originating role. When you select Copy Role, you will first be able to set the name of the new role (see Figure 2.4). After you have entered the name, click OK to proceed.

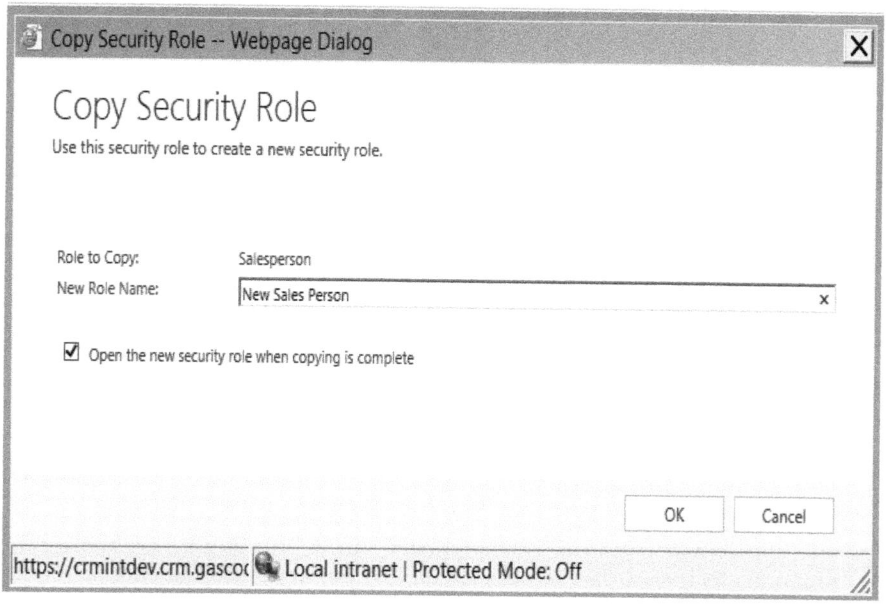

Figure 2.4: Naming your newly copied role.

Once the copying is complete, the new security role will be displayed. The main page allows you to determine which business unit the role is part of. You will also see many tabs, which allow you to configure access to each individual entity and functional area within CRM. The colored circles on each of the tabs define the access level for each privilege.

Access levels determine how deep or high in the organizational business unit hierarchy a particular user can perform a specified privilege. Figure 2.5 shows the main page of the new security role's settings, with a color-coded key of privileges at the bottom.

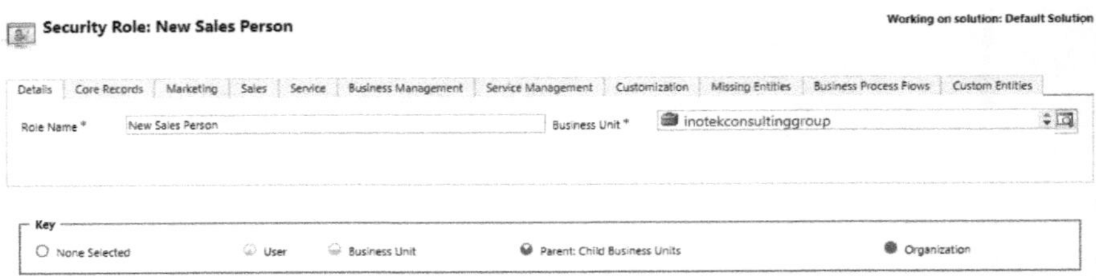

Figure 2.5: Many tabs are available for role security settings—don't miss the key at the bottom that shows what the symbols mean.

The options for these privileges are as follows.

- None Selected—No access to that entity.
- User—Only records that the user owns.
- Business Unit—Can access records within the business unit to which the user is assigned.
- Parent: Child Business Units—Can access any record in the user's business unit and any records in business units that have been configured as children of the user's business unit.
- Organization—Full access across all business units.

Each tab contains entities specific to that tab as well. For example, in the Core Records tab, entities like Accounts, Contacts, Activity, and many other entities are found. In the Marketing tab, Campaigns & Marketing Lists are found. An example of a role's privilege configuration for the Core Records tab is shown in Figure 2.6.

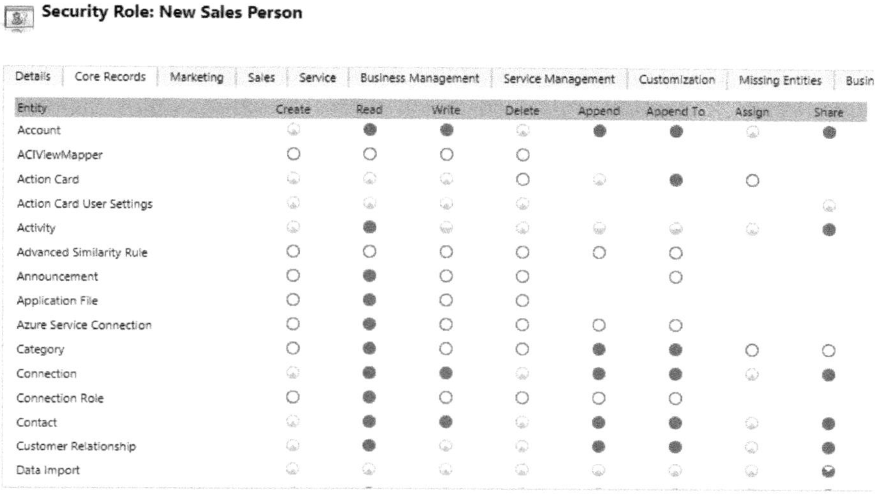

Figure 2.6: The security settings on core records.

Beyond granting access to an entity and determining the level of access, each tab also contains specific functions (located at the bottom of each tab) that can be turned on or off by role. Referring again to the Core Records tab (but scrolling to the very bottom), Figure 2.7 shows the functions that can be turned on or off.

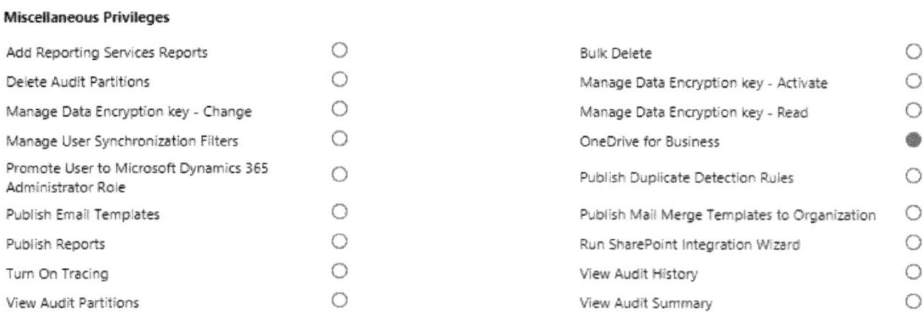

Figure 2.7: Look at the bottom of each tab for additional privileges—you will have to scroll down

Users

We will now look at creating user records. A *user* is any person who works for a business unit who uses Microsoft Dynamics 365. Each user must have their own user account, and will be associated with only one business unit at a time. This association controls which customer data the user will have access to. Included in the user's account is information such as the user's telephone numbers, email address, and a link to the user's manager.

Each user has privileges and rights to manage their own personal settings. Each user corresponds to a user in the Dynamics 365 Office Admin Portal. The first step to adding a new user is to ensure you have created your user in this Portal. One easy way to access this is by going into your Settings and clicking Users. From here, you will see a list of current users. Clicking the "New" button will display the message shown in Figure 2.8, along with a link to the Admin Portal.

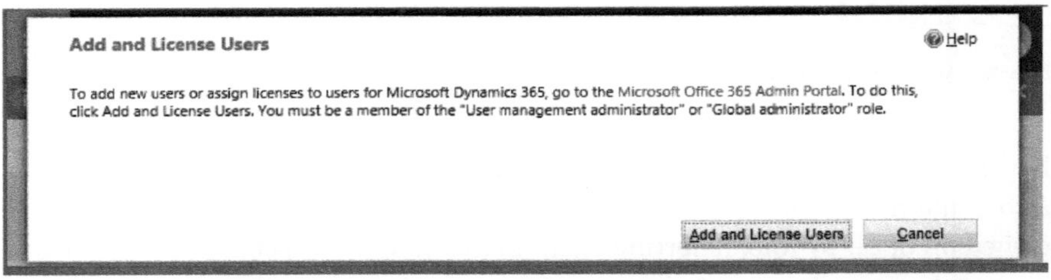

Figure 2.8: Add a new user from within CRM.

Clicking the "Add and License Users" button will take you to the root Office 365 portal. Click on the "Admin" button, which will get you to a dashboard where you can add a new user (see Figure 2.9).

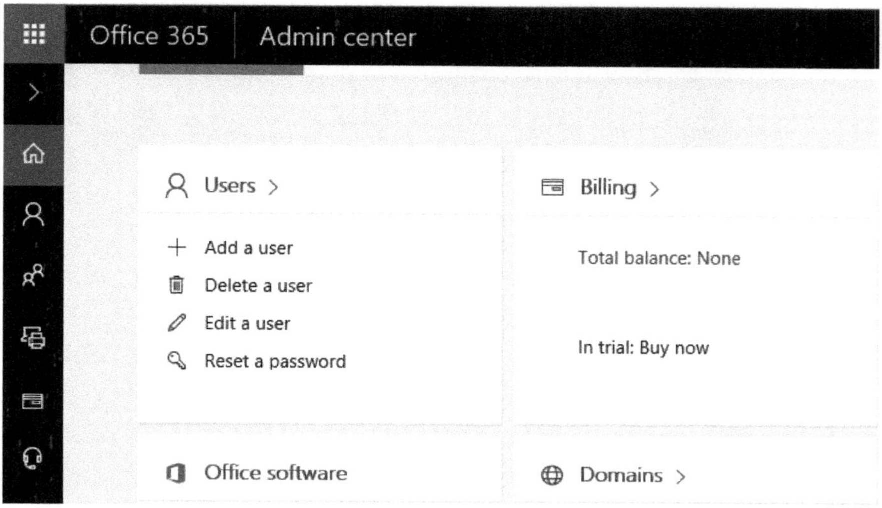

Figure 2.9: The Office 365 Admin Portal is where new users are created for Dynamics 365.

Clicking the Add a user link will open the window shown in Figure 2.10. Most of the information that you will need to enter is obvious, but the "Product licenses" section is where you purchase the license, and in turn make the user available for login and assignment within Dynamics.

Each user you create will cost more in your CRM subscription. During the 30-day trial, there is no charge.

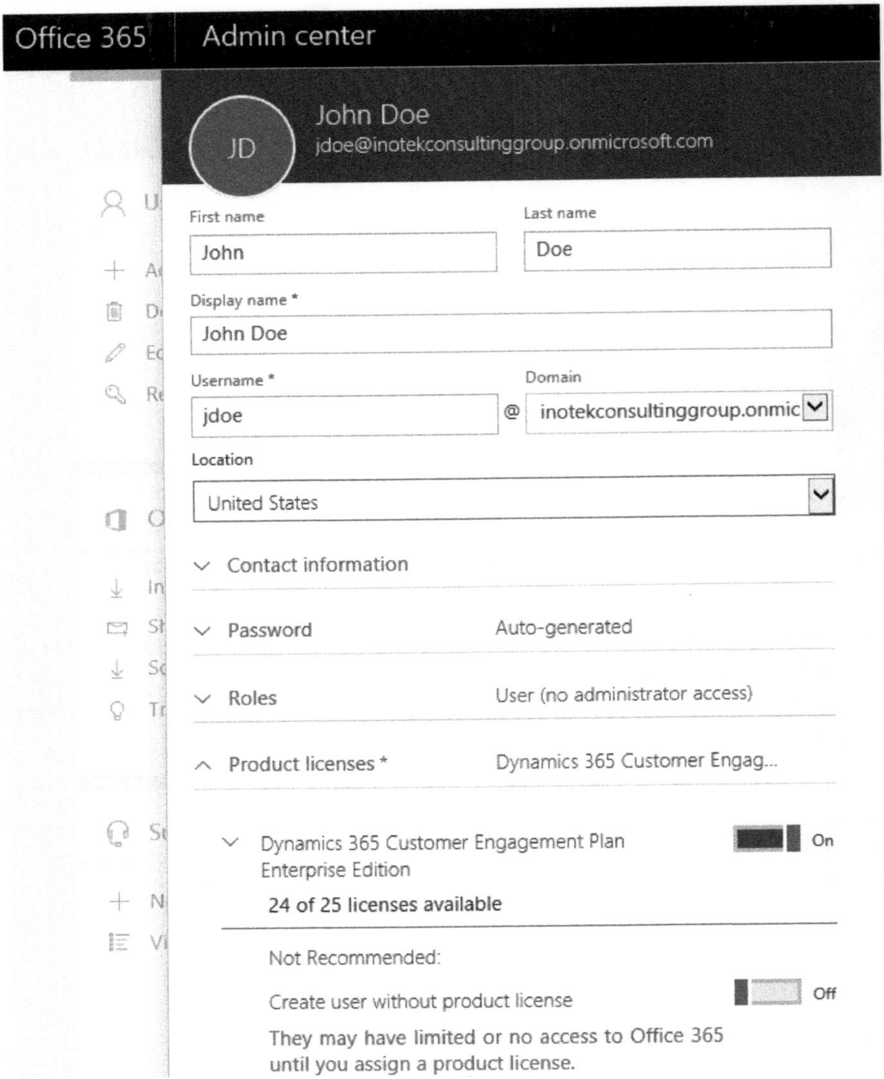

Figure 2.10: Adding the user in the portal and setting the appropriate license.

Once you save the record, you can now go back into CRM and you will see the user that you have created available in the Enabled Users list (see Figure 2.11).

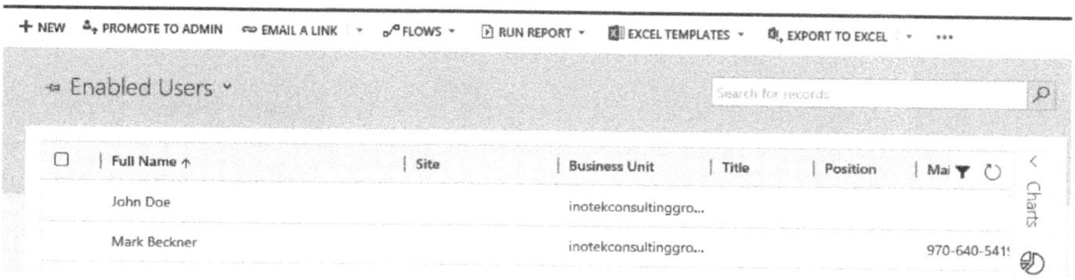

Figure 2.11: The user now appears in the Dynamics list of users.

When you create a user in Dynamics 365, you must assign the user to at least one security role. Even if the user is part of a team that has assigned roles, the user should be assigned to a specific role. You can assign roles to a user by opening the user record and clicking on the "Manage Roles" button in the toolbar (see Figure 2.12).

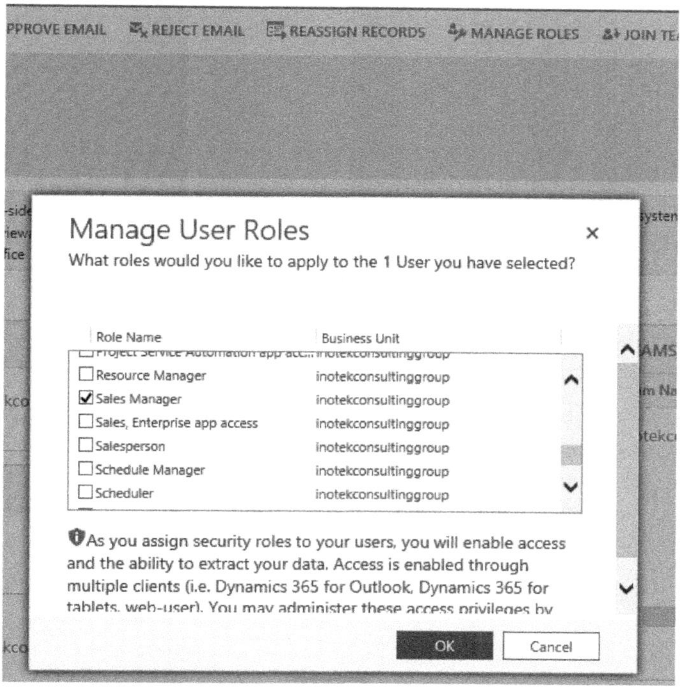

Figure 2.12: Assigning a role (or roles) to a user.

Teams

A *team* is essentially a collection of users. By creating a team, a record can be owned by a team, thus allowing any user on the team access to the specific entity and/or data records assigned to the team. To setup a Team, return to the main Security screen and click on Teams. The window in Figure 2.13 will open. By default, CRM is configured with a team that has the same name as the organization.

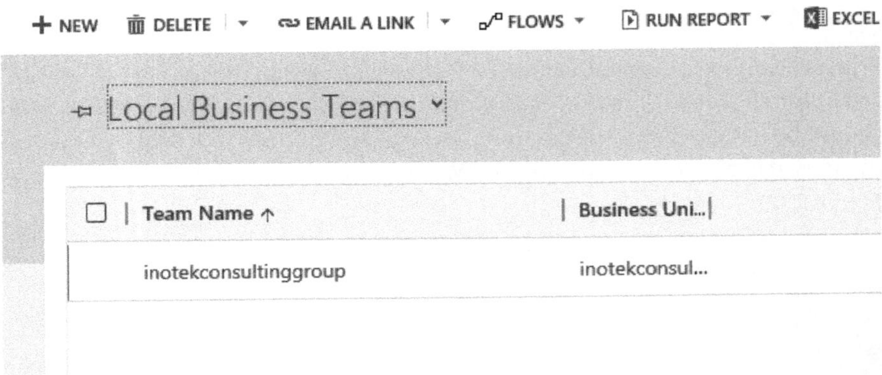

Figure 2.13: The list of teams.

Click the "New" button and the Team form will load. The following list shows the mandatory fields. Once these fields are completed, click the "Save" button. Members can now be added to the team using the section on the right (use the plus sign in the right-hand corner of the control area), as shown in Figure 2.14.

- Team Name—This is the name of the team.
- Business Unit—This is the business unit that the team is part of, and will inherit base permissions from.
- Administrator—This is the owner of the team record.
- Team Type—There are two options. The first is Owner, which denotes that this team can *own* records. The second is Access, which allows them *access* to records via sharing.

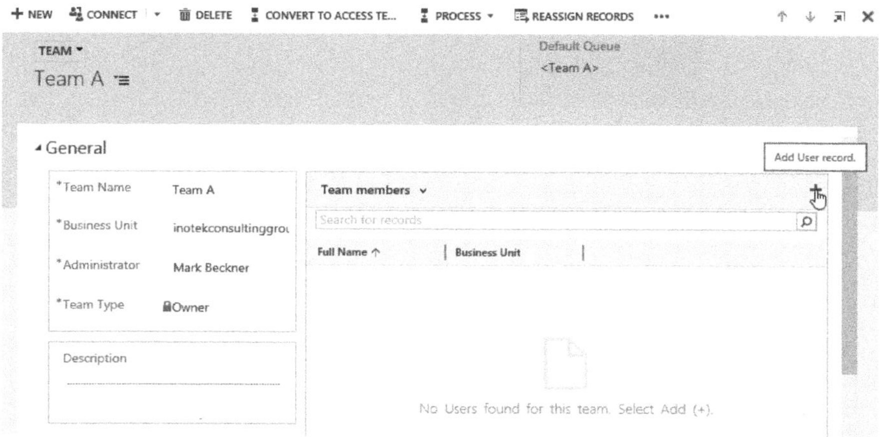

Figure 2.14: Adding a user to a team (note the little plus sign in the corner).

After saving the new Team, the Add User record icon (the plus sign) displays. Click the plus sign and add all or part of a name to the search field. Click the search button (the magnifying glass) and a list of existing users will be displayed. Selecting the user from the results list will cause it to be added to the team. You can add as many users as you wish to a team. You can also join teams from within a user record by clicking the "Join Team" button on a User form. When this button on the toolbar is clicked, a Look Up Records screen will be opened (see Figure 2.15). Select the team from the list and click Select, then click Add to close the screen.

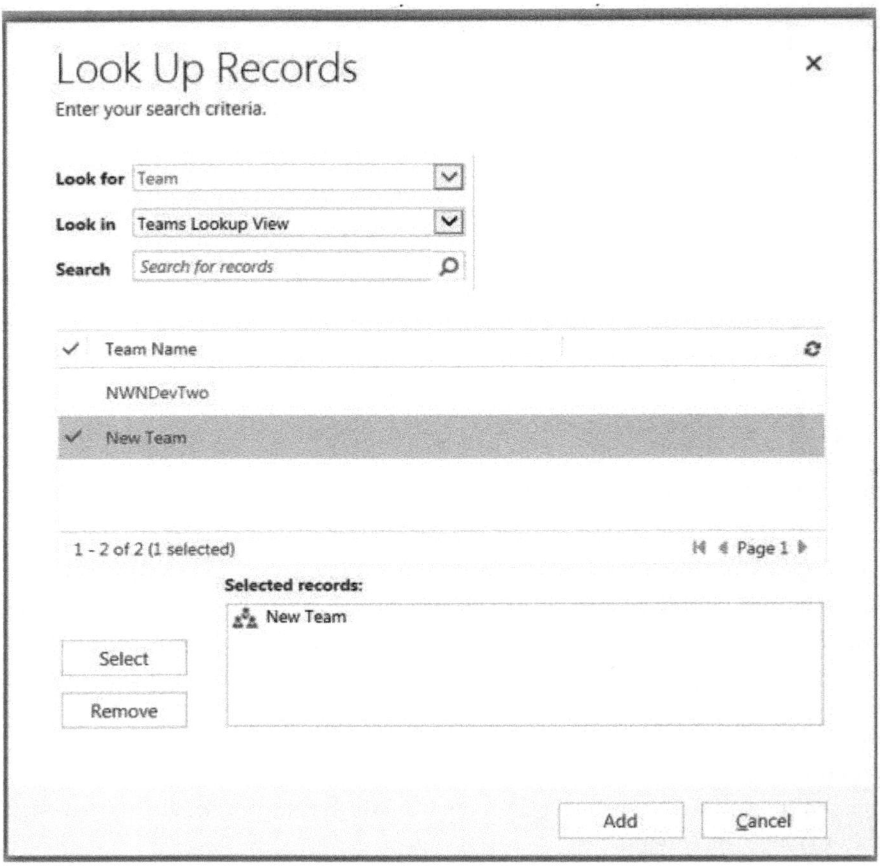

Figure 2.15: Adding a team reference to a user record from the User form.

Note that records can be owned by teams or by individual users.

Field Level Security

Using *field level security* allows restrictions on accessing and updating attributes on an entity. *Record-level permissions* are granted at the entity level, but you may have certain fields associated with an entity that contain data that is more sensitive than the other fields. For these situations, you use field level security to control access to specific fields.

The scope of field level security is organization wide and applies to all data access requests, including the following:

- Data access requests from within a client application, such as web browser, mobile client, or Microsoft Dynamics 365 for Outlook.
- Web service calls using the Microsoft Dynamics 365 SDK (for use in plugins, custom workflow activities, and custom code).
- Reporting (using Filtered Views).

Before you can define security on specific fields, you must open your entity (via Settings, Customization, Customize the System, select the entity, select the Fields item, select the field) and set the Field Security attribute to Enable, as shown in Figure 2.16. Once Field Security is enabled on an entity, and the customization has been published, the attribute will appear for that entity and can be modified.

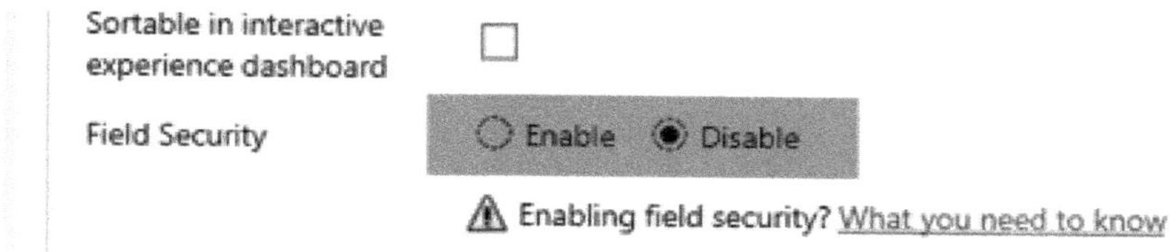

Figure 2.16: You must open each entity where you want Field Security to be enabled and then publish the changes.

To set up field level security, you can begin by clicking on the Field Security Profiles link from the main Security section (accessible from the Settings menu on your toolbar). Figure 2.17 shows where this link can be found.

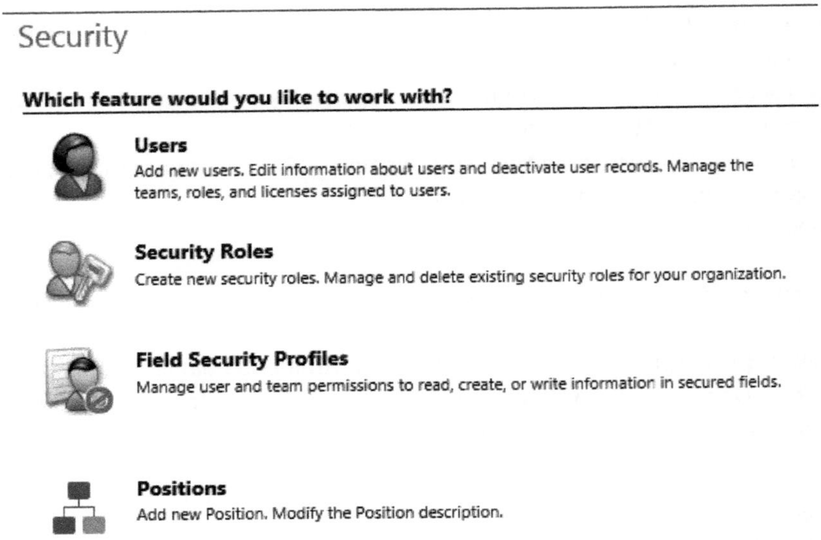

Figure 2.17: Accessing Field Security Profiles from the main Security tab.

Clicking Field Security Profiles opens a new window where you can then click on New to create a new profile. You must begin by entering a Name, and then save the record. Once the record is saved, teams and users can be assigned to this profile. To update the attributes identified as having field security enabled, select the Field Permissions option from the left hand navigation, as shown in Figure 2.18.

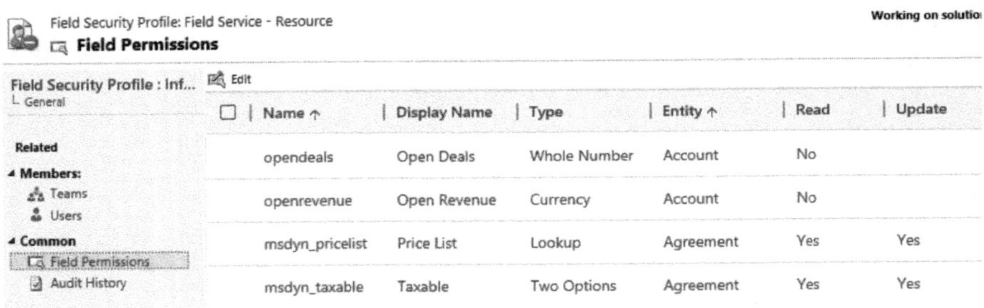

Figure 2.18: Setting individual permissions at the field level within an entity.

Select the attribute that needs security assigned, and click the “Edit” button. You can update the Read, Update, and Create properties as appropriate, and then click OK (see Figure 2.19).

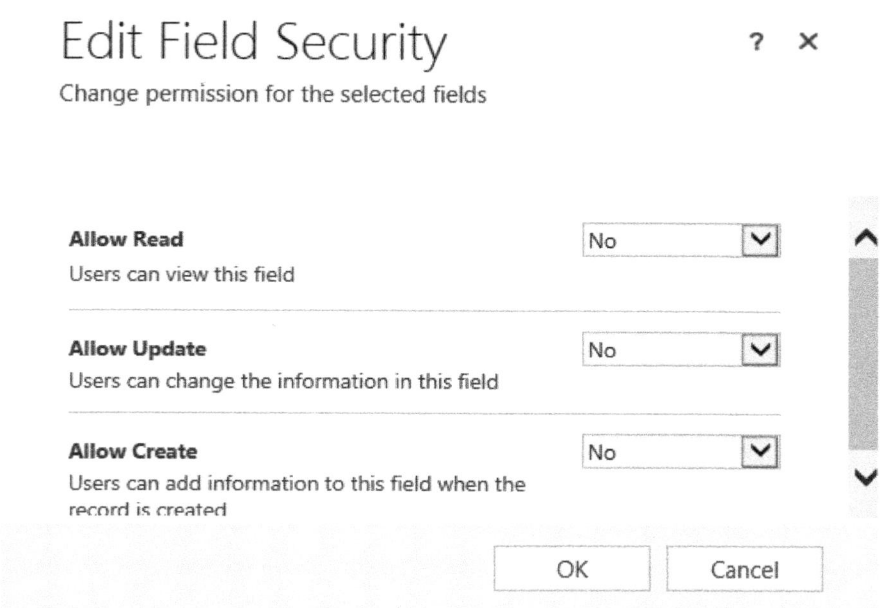

Figure 2.19: Editing the access on a field.

Once Field Security is enabled on a field and a Field Security Profile restricts the attribute, it will now function according to how the profile has been configured. For example, Field Security was enabled on the Gas Availability attribute shown in Figure 2.20. This attribute was added to a Field Security Profile. Notice the little key icon next to the attribute name—this means it has security configured at this level.

Figure 2.20: A key icon will appear next to the field once Field Level security has been added.

Summary

Security can be a little overwhelming at first, but as this chapter has demonstrated, it can be markedly simplified if you understand what is available to you and how you can use it. Some important concepts around security include the following items:

- To add users in CRM, they must first be setup within your Office 365 Portal.
- CRM comes preconfigured with many default roles that can be utilized in the initial setup.
- The main two roles to assign to users that are to be able to configure CRM are:
    - System Administrator
    - System Customizer
- Beyond a User, CRM has the concept of a Team, a group of users with the same job function. Like users, security roles can be assigned to teams.
- When CRM is installed, a base Business Unit is configured. Additional Business Units can be added and assigned to a hierarchy that should match the business structure for the organization utilizing CRM.
- Business Units are leveraged in assigning and controlling access to records. This is done as part of the security roles setup.

Chapter 3
Customizations

Dynamics 365 is inherently customizable. You can modify and create just about anything you need within the environment in order to accurately model your organization and work with the underlying data. This chapter will introduce how to modify entities and add fields of various types. It will also introduce basic concepts around entities, JavaScript, and views.

Adding Fields

The easiest way to jump into customizing your CRM environment is to add a *field* to a form. We will create several types of fields on the existing Account entity to demonstrate what options are available to you. In order to add these fields, begin by clicking on Customize the System from the main Settings menu. In the window that opens, expand Entities, then expand Account and click on the Forms tab (see Figure 3.1). From here, you will be presented with a list of all of the existing forms that are available for the Account entity. Click on the form that is simply titled "Account," this is the main default form that will open in an out-of-the-box Dynamics 365 environment.

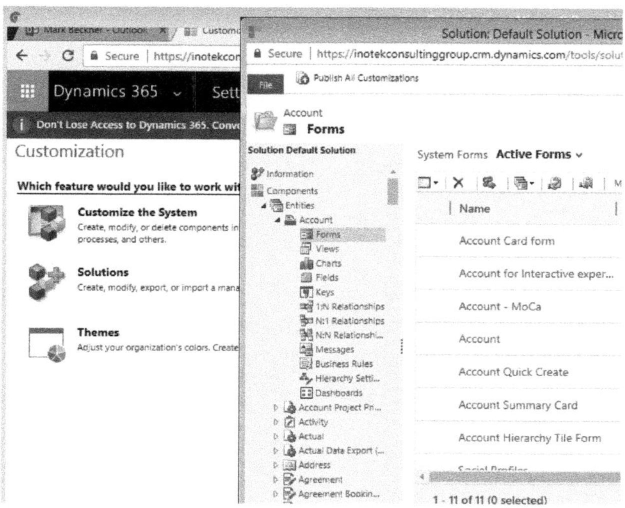

Figure 3.1: The Forms tab where you can access all forms on an entity.

DOI 10.1515/9781547400065-003

With the form now open in an editable format, you will see what is called the Field Explorer open on the right hand side (refer to Figure 3.2). This shows all the fields that are available on the Account entity that have not already been added to the form. You can drag a field from this window and drop it on the main form area, or you can create a new field to add to the form.

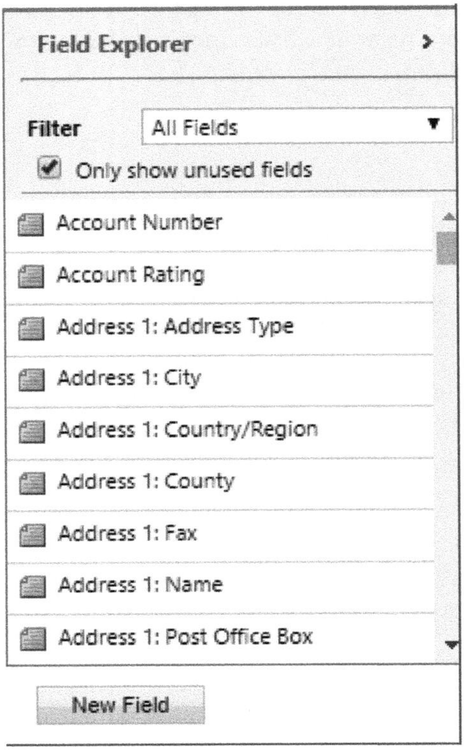

Figure 3.2: The Field Explorer shows fields that can be added to the form.

The first field we will create is a multi-line text field. To add a new field, click on the button labeled New Field, which is at the bottom of the Field Explorer. In the window that opens, enter the following information (use Figure 3.3 for reference).

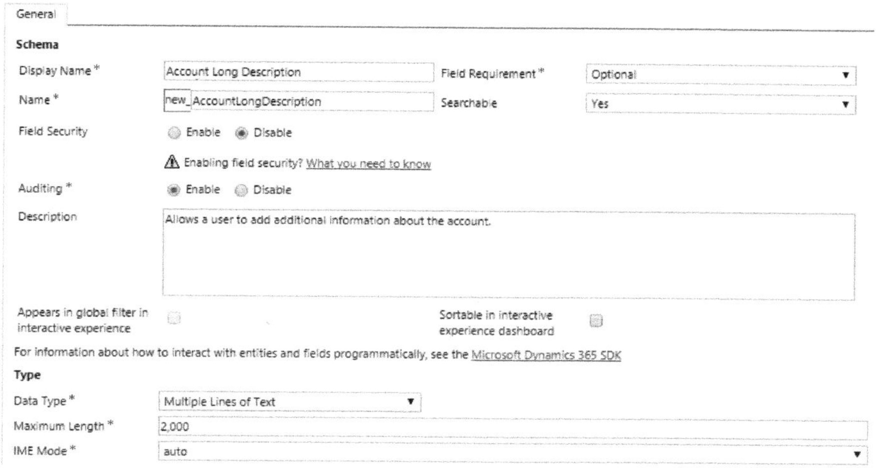

Figure 3.3: Creating a multi-line text field.

- *Display Name*—This will be the text that shows in the label of the field. You can override this once it is added to a form, if needed.
- *Field Requirement*—You can set your field to be optional or required. If required, the user will not be able to save the form without populating this field.
- *Name*—This is the name that will be given to the field at the schema level. The schema is the online database name, and if you program a plugin or other code and want to reference the field, you would use its schema name. If you wish to have a different prefix on your field names, you will need to create a new publisher (see Chapter 8).
- *Searchable*—Here you indicate whether the field will be included in searches. The search that this refers to is located in the upper right portion of the window when you are on the main entity view. Figure 3.4 shows the search window for Accounts. This is based on the default Account view that opens when clicking Sales on the main menu and selecting Accounts.

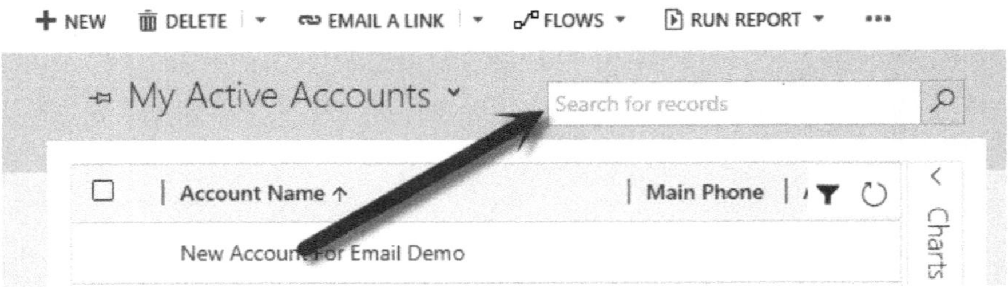

Figure 3.4: If a field is searchable it will be included in record searches.

- *Field Security*—You can set access to a field based on user permissions. See the section in Chapter 2 on Field Level Security for more detail on this topic.
- *Auditing*—You can enable or disable auditing on a field (see Chapter 1). It is best to keep this setting always enabled so you have the option to track changes easily.
- *Description*—It is good practice to put something in this property for fields that may not have an obvious usage. Remember that there will be multiple people who configure any system and this field provides an opportunity to easily document the purpose of a field.
- There are two fields with long titles. One is: *Appears in global filter in interactive experience*; and the other is: *Sortable in interactive experience dashboard.* When enabled, these will allow users to access them and work with the field in charts for dashboards.
- *Data Type*—There are many field types that you can select, including option sets, Boolean values, images, and lookups to other entities. We'll cover several of these in this chapter, but for the current example, you'll want to select the Multiple Lines of Text option.
- *Maximum Length*—Note that this property is specific to the Multiple Lines of Text data type option. Other data types will have different properties to set here.
- *IME Mode*—Use this field when you plan to enter foreign characters, such as Chinese language symbols. This field is generally set to auto.

When you have completed all details, you can click the Save and Close button at the top of the window. This will create the field and take you back to the main editing window. To show your new field in the Field Explorer, you will likely need to refresh your screen. If you haven't made any edits to the main form, you can simply refresh the window by pressing F5. If you have made edits, you'll want to save what you've done and then refresh or re-open the window. When you do, you will see your new field in the Field Explorer list.

You can now add our new field just about anywhere on the form. In the example shown in Figure 3.5, the field has been dropped right under the Account Name field on the form.

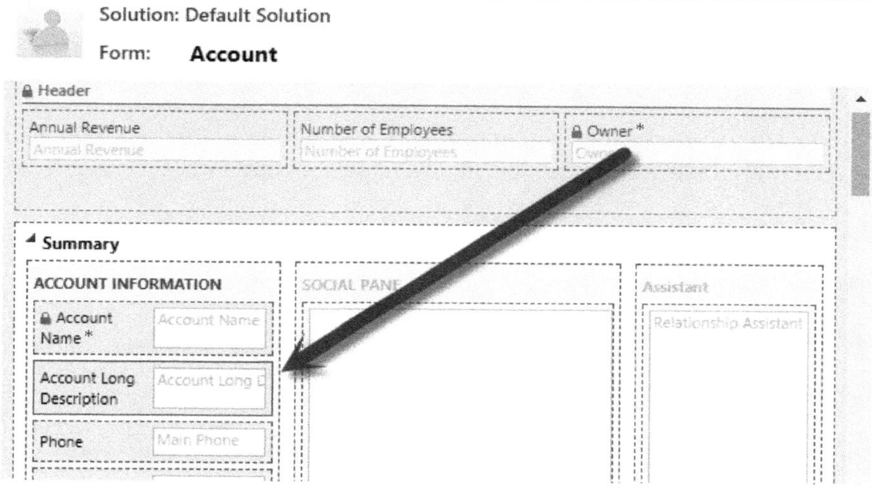

Figure 3.5: Dragging the new field onto the form.

In this case, you've added the field, but the real estate is very limited. For a field that is a simple drop down or short text field, this placement would be adequate, but for a field that has been enabled for multiple lines of text, you'll want to place it somewhere else on the form to allow plenty of room to enter information. To do this, you can create a new section on your form and drag this field into that section. To do so, follow these steps:

1. Scroll down to below the main account summary area and click on Project Price Lists (or some other area). This will give context to where you wish to add the new section.
2. Click on the Insert tab at the top of your editable form and click the One Column tab (shown in Figure 3.6). This will create a new tab that has one large column in it.

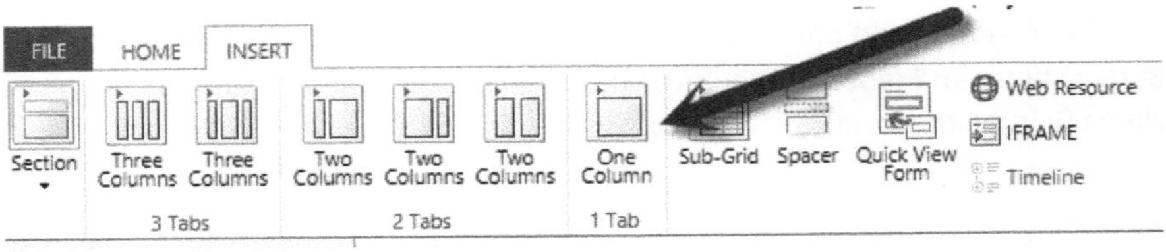

Figure 3.6: Adding a single column tab to separate data on your form.

3. The tab will automatically populate with a section in it. You can drop your multi-line field into this section. Figure 3.7 demonstrates what this now looks like.

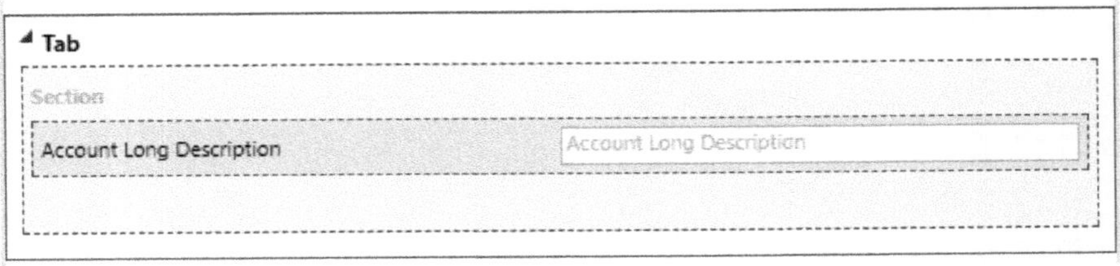

Figure 3.7: Moving the new multi-line field to the new tab.

4. You can change several things about your tab by clicking on the tab and then selecting the Change Properties button on the toolbar on the Home menu. Figure 3.8 shows where this button is and the screen that opens when you click this button. In this example, the label on the tab has been altered to be more descriptive and give instructions to the user.

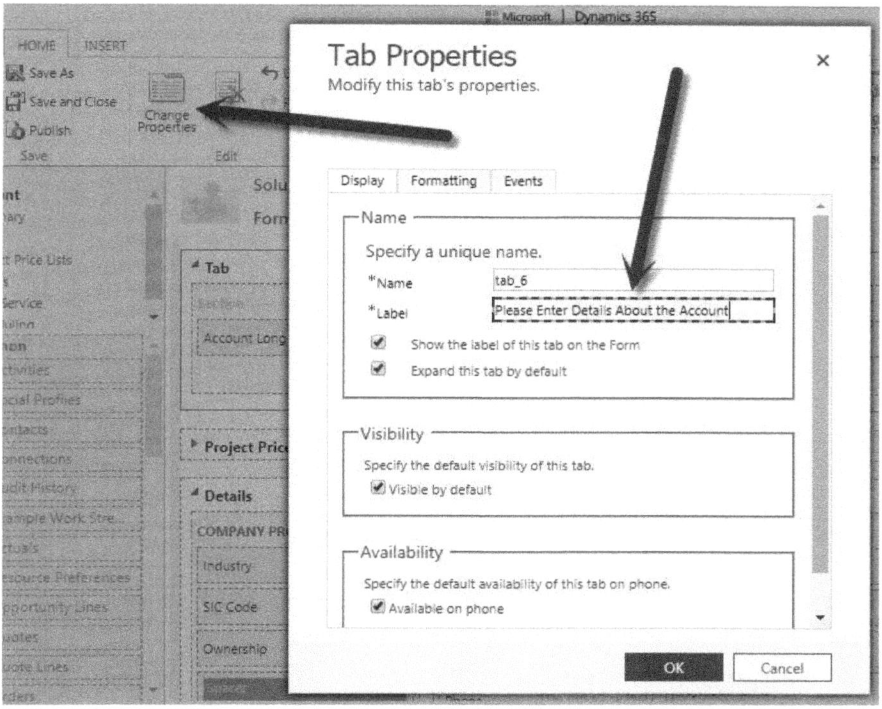

Figure 3.8: Accessing the properties of the new tab to add descriptive labeling.

Publishing Your Changes

You have created several fields, but they are not yet available to use or view on the actual account form. In order to make your new fields available, you must publish them. You have two choices when you publish. You can publish only the changes for the entity or component you have been working on, or you can publish changes for all entities and components. In this case, you just want to publish the changes that you've added to the Account entity. To do this, click back on the Entities tab in your solution and select the checkbox next to the Account entity in the list (see Figure 3.9). Then, with the entity selected, click the Publish button at the top of the view.

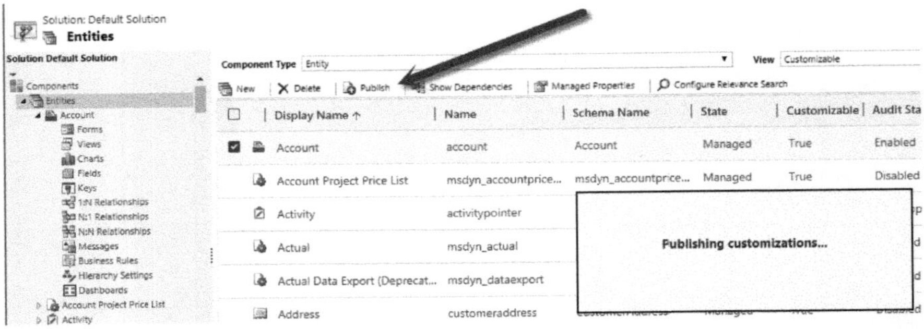

Figure 3.9: Publishing the changes.

Once the Publishing customizations message disappears, you'll be able to see your changes on your main account form. You may want to refresh your browser after a publish to ensure that you have access to the latest changes. Figure 3.10 shows the long description field that was added in its own section.

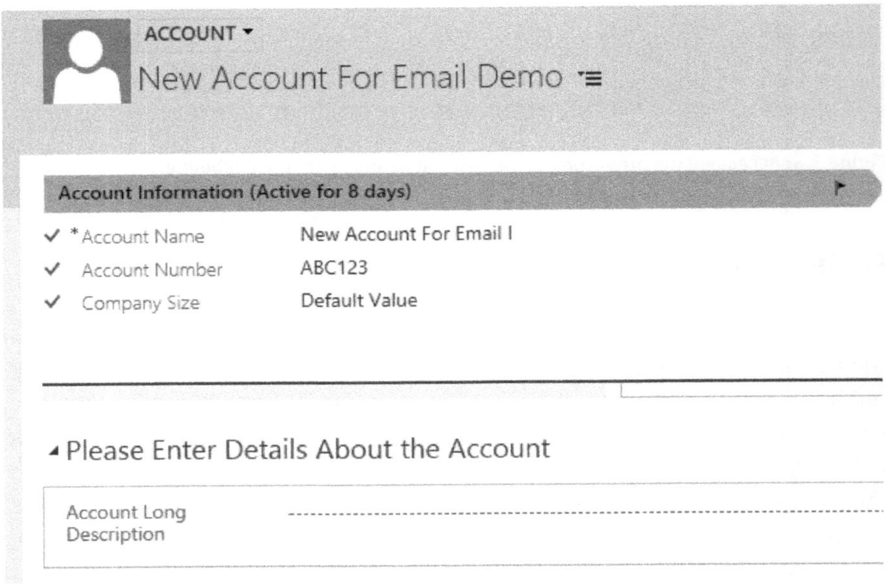

Figure 3.10: The new field in action on the form.

Option Sets

Now let's create a new field. This new field will be an option set, which will provide a drop-down from which the user will be able to select a value. To create an option set, add a new field just like before. In the Type section of your new field, select Option Set. A number of properties appear that can now be configured. These include the following:

- *Field Type*—Simple or Calculated. A *simple* option set contains static text values that you can type in now. A *calculated* option set allows you to create rules for what data will appear in the option set. Figure 3.11 shows the window that opens where you can configure these rules. In this example, the option set value will have Option A shown in it if the condition is met. You can build this out with as many conditions as needed. For this example, we will use a Simple option set.

You also have the option of using JavaScript in your form to show and hide values in your Option Set list.

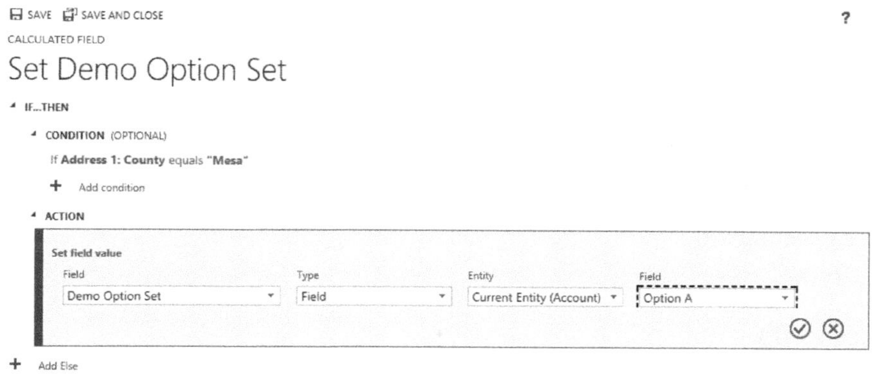

Figure 3.11: An option set with calculated values.

- *Use Existing Option Set*—If you have a set of values that will appear in your option set that exist in other option sets either on this entity or other entities, you can choose to have all entities share a single option set value list. In this example, we'll simply configure this option set with its own unique values that are not shared, so set this value to No.

- *Default Value*—This will show as Unassigned Value by default, which means when a form loads, no value will be selected. If you wish to have this default to one of the options that you are going to create, first create the option, then come back to this field and set it to that value.
- *Options*—Here you can add in your list of options. Click the plus sign to add a value, and edit the label and description in the right-hand column. Figure 3.12 shows several values being added to the list.

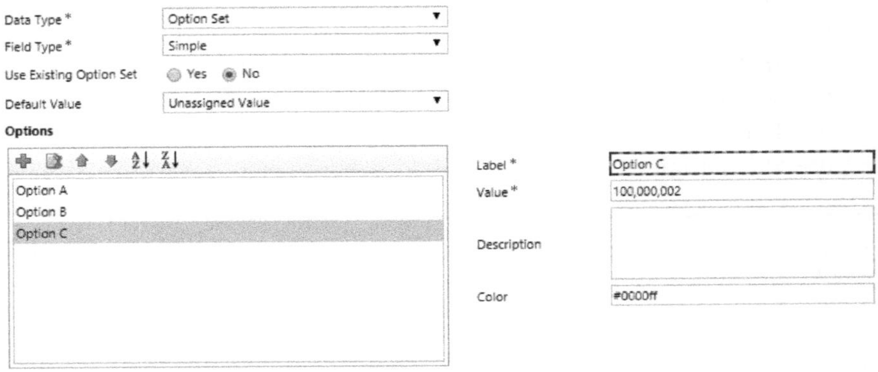

Figure 3.12: An option set with static values.

When you are done configuring your new option set, click the Save and Close button and drag-and-drop the new field on to your form. Save your edited form, publish it, and you will see your option set on your user form (see Figure 3.15 in the next section).

Lookups

A *lookup* field allows you to add a relationship to another record. A lookup can be on the same entity type (for example, an account may have a parent account) or it can be on a record of a different type (a contact can be related to an account). In order to create a lookup, begin by creating a new field. You can create a new field straight off the form, as demonstrated earlier in this chapter, or you can create it from the Fields tab. Go ahead and click the New button from the Fields menu (Figure 3.13 shows where to find this button).

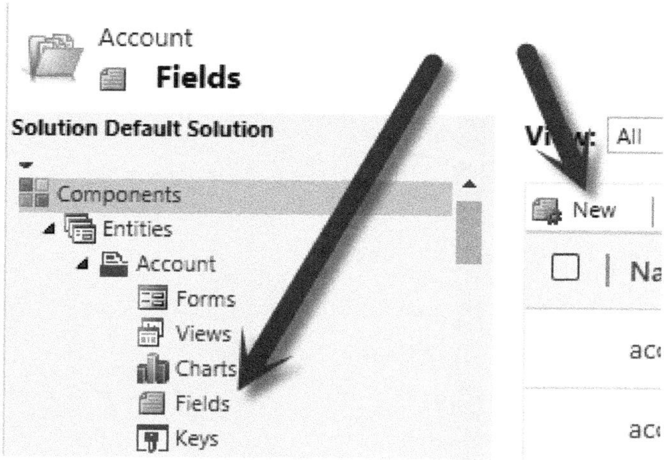

Figure 3.13: Creating a new field from the Fields tab.

In the field configuration screen that opens, select the Lookup value from the Data Type drop down. Doing this will then allow you to select Account from the Target Record Type. You can create a lookup on any entity type (see Figure 3.14). The last property that you can set on a lookup is the Relationship Type (see the next section in this chapter for details on relationships).

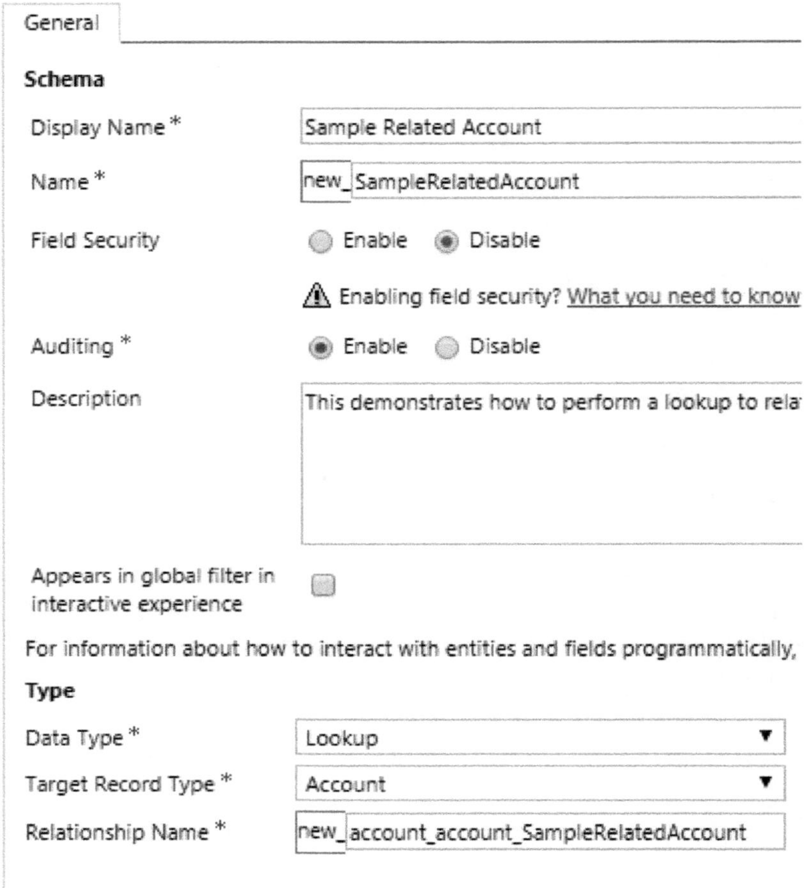

Figure 3.14: Creating a lookup field.

Click the Save and Close button on your new lookup field and then add it to your form. To do this, you'll have to open your form back up (if you created your field from the Field tab) and then drag it from the Field Explorer and drop it on your form. Figure 3.15 shows that this new lookup field (called Sample Related Account) has been dropped right below the Demo Option Set field created in the previous section.

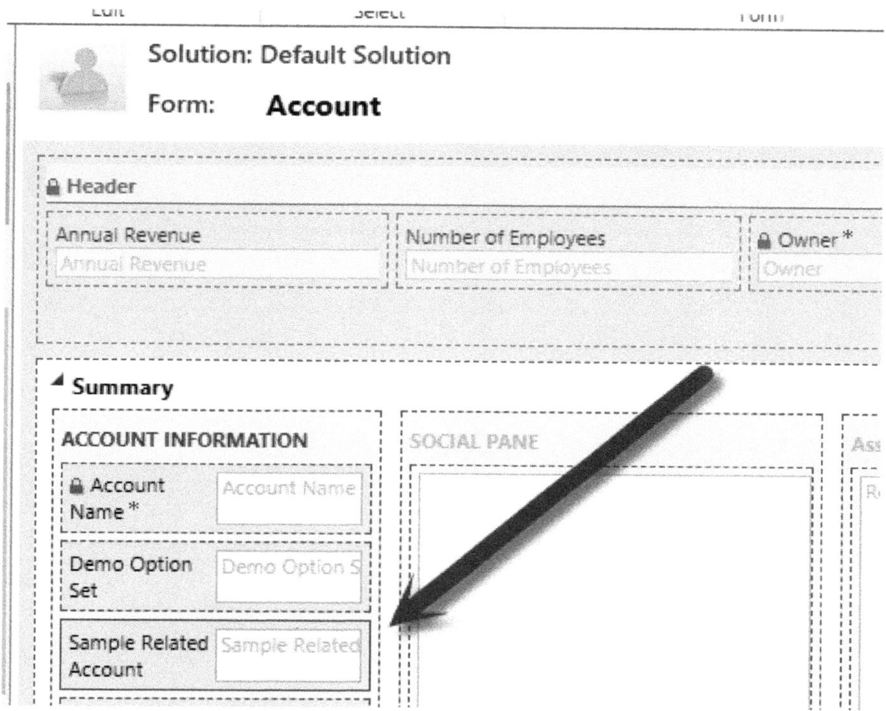

Figure 3.15: Adding the lookup to the form.

Once you have the field placed, publish your entity. The lookup field will now display on your form (Figure 3.16). When you click the field, you will get the standard lookup search functionality that will allow you to find the specific record you are after.

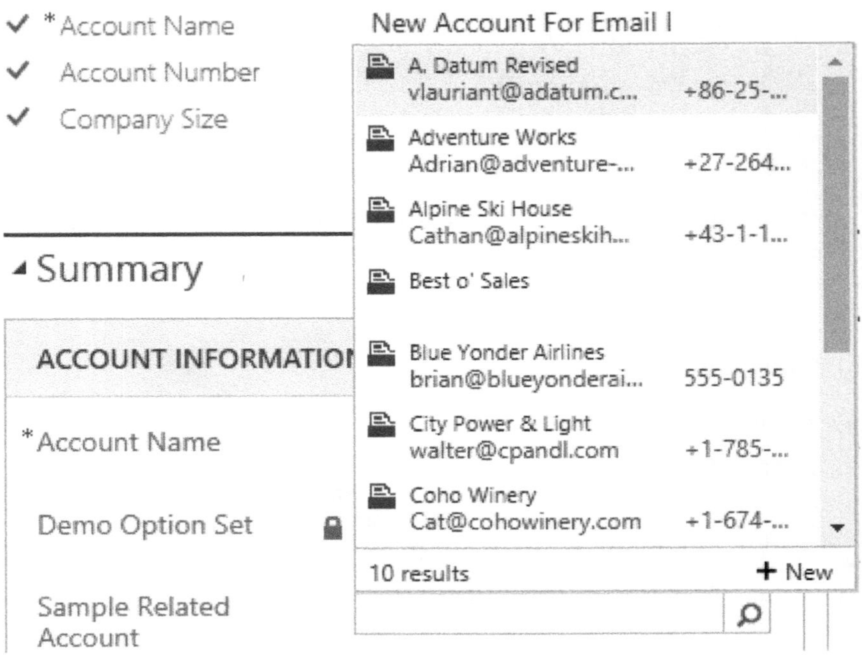

Figure 3.16: The lookup in action on the published form.

Relationships

Every lookup that you add to your form will create a *relationship*. When you add a lookup and let the relationship be automatically created, the system will create a one-to-many relationship. To see your lookup relationship, click on the 1:N Relationships tab under Account and then sort by the Related Entity field in the view on the right (see Figure 3.17).

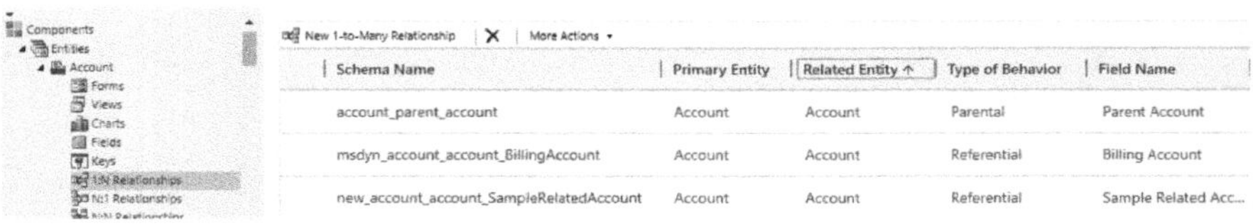

Figure 3.17: The Relationships tab.

By defining your own relationship type, you can allow users to easily relate and view data across entities from a single form. For example, if you wish to allow users to create a single "Related Account," then you can use a lookup field as described above. If you wish to allow users to create many "Related Accounts," you can create a many-to-many relationship and then add that relationship to your form. To do this, click on the N:N Relationships tab under Account and click the New Many-to-Many Relationships button. The screen in Figure 3.18 will open. Set this relationship as shown and click the Save and Close button.

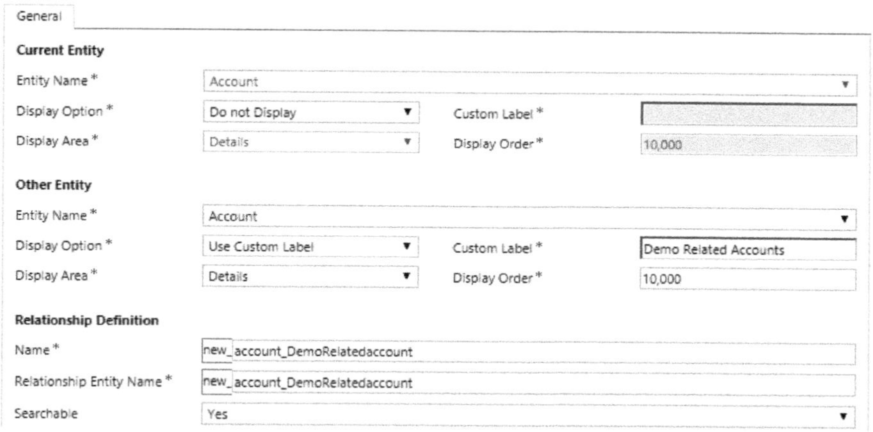

Figure 3.18: Creating a new many-to-many relationship.

You can access your relationships on the form editor by clicking on Navigation in the main ribbon (see Figure 3.19).

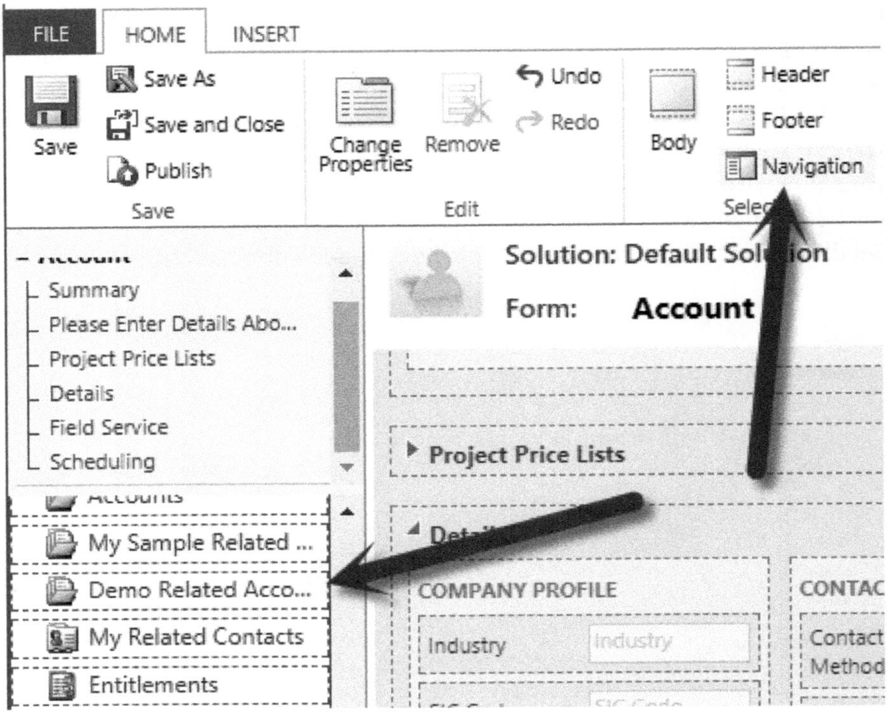

Figure 3.19: The relationship will appear in the navigation section of the form editor.

Once you have published your customizations, your new relationship will be available from the main menu on your specific record. To access it on an account, for example, click on Sales, then select Accounts and open a specific account. Once that account is open, click on the little downward arrow at the very top and you will see your custom relationship available from this area (Figure 3.20).

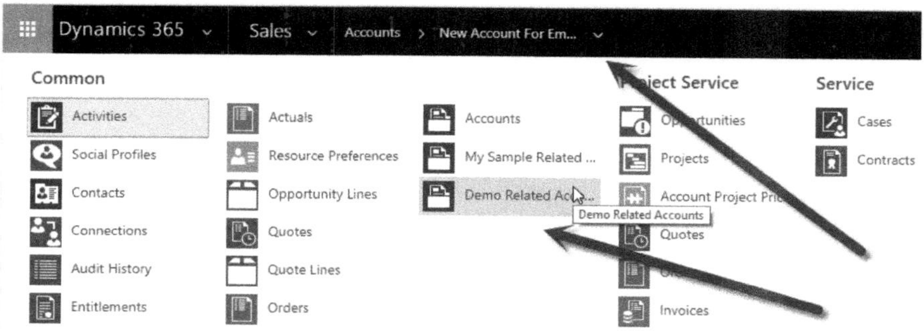

Figure 3.20: Accessing relationships through the main menu navigation.

Creating New Entities

Customizing existing entities is a great skill to have, but to really leverage the power of Dynamics, you'll want to know how to create custom entities. Custom entities allow you to quickly model unique aspects of your business, and to rapidly develop forms, fields, relationships, and other related components that will support this unique model. For example, let's say that you had a business that managed political donations. You may track your donors as contacts or accounts, but you also want to keep track of what organizations these individuals are associated with. Individuals could be associated with many political groups, and you want to track these political groups as a new entity, with fields that are specific to politics, cash contributions, etc. We'll use this example to build out a custom entity, and show how to relate this entity to a Contact.

To create a new entity, click on Entities in your default solution, and click the New button. A small menu will open where you can select Entity (see Figure 3.21). You can create many other component types from this menu as well.

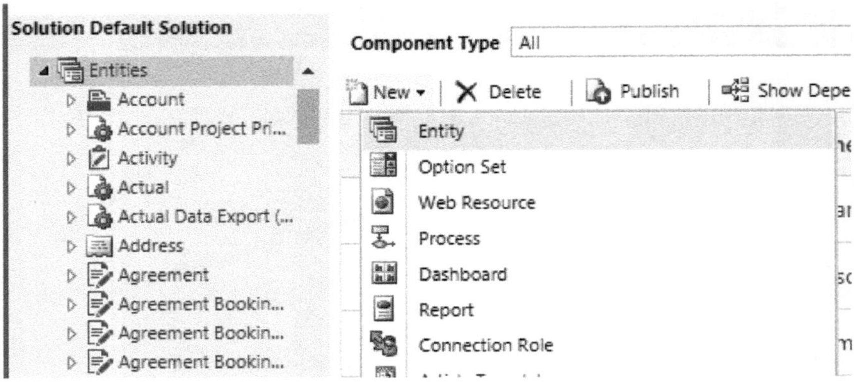

Figure 3.21: Creating a new entity.

When you click the Entity button, a new window will open with a large number of fields that can be configured. We'll look only at the significant options, but you will want to explore and become familiar with all of them. The main section is Entity Definition (shown in Figure 3.22), which is where you fill out the core information about your entity. The Display Name property is what the entity will be called. For this example, name it Political Group. The Plural Name field is used to let CRM know how to refer to more than one of these. Political Groups would be the value for this example, but if you had an entity called Property, the plural value would need to be entered as Properties.

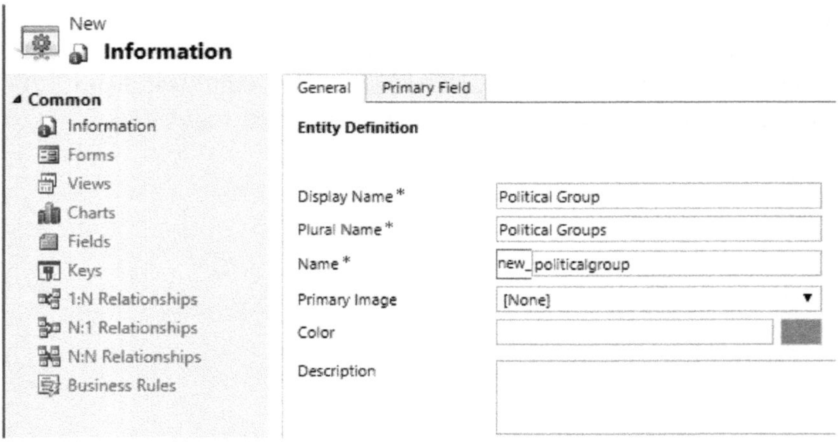

Figure 3.22: The initial configuration of the entity.

Click the Primary Field tab. By default, the entity will require a Name to be entered (see Figure 3.23). You could change this to something else—like a number or other value required to create the record.

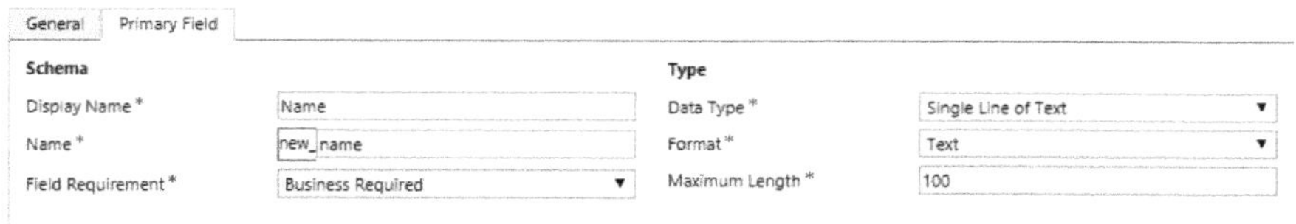

Figure 3.23: Every entity must have at least one required field.

Once you have the basic information entered, click the Save button to create the entity. This will cause a pop up to display (Figure 3.24). It can take a few seconds for this to complete, and you won't be able to do anything within the browser until this has finished.

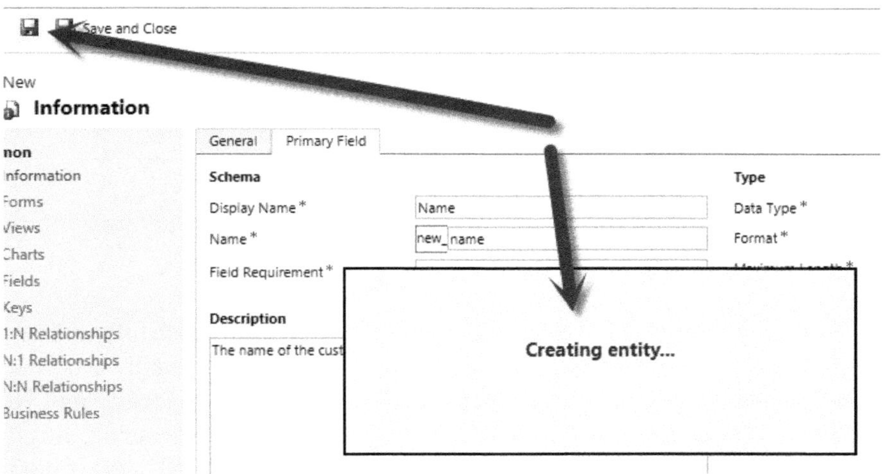

Figure 3.24: Creating the entity.

When the creation of the entity is complete, you can close the creation screen, navigate to the Political Groups Entity in the Default Solution Explorer, and click on the Forms tab. When an entity is first created, there are only three forms. The form that has a Form Type of Main is what will display by default. Click on this main form to open it. There will only be two fields on your form—the Name (which was defined as your Primary Field) and the Owner (every record will always have an Owner on it—this is the user that created the record). There are a few additional fields that can be added to the form, such as Status and Modified By, but overall there is very little on your new form and entity. Figure 3.25 shows this new entity's editable main form.

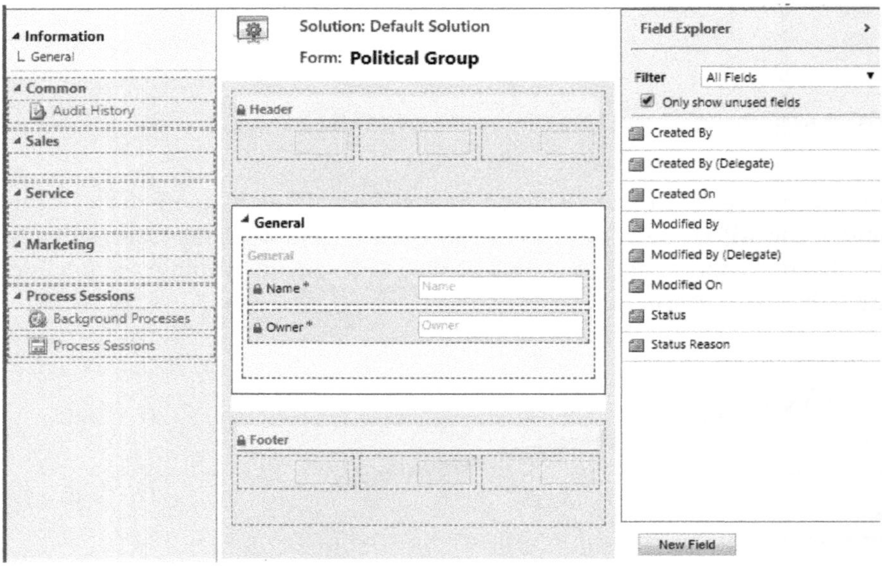

Figure 3.25: The new entity's editable form

Click the Save button on the form editor, and then click the Publish button. Your form will now be published—but unless you specified where it would be located in the navigation menus, it will only be available through Advanced Find. To specify from where the form will be accessible, you can click on the main Political Group name in the Solution explorer, and on the General tab, scroll to the section titled "Areas that display this entity" (as shown in Figure 3.26). Here you can determine where your entity will appear once it is published. In this example, it will be available from the Sales navigation menu.

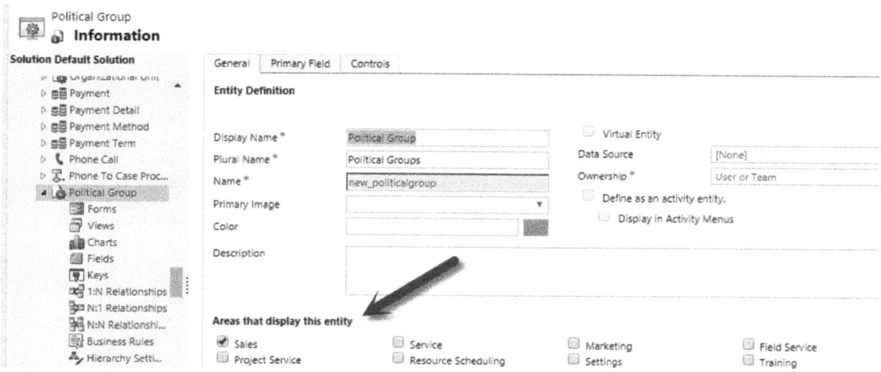

Figure 3.26: You can determine where in the navigation this new entity will appear.

When your entity has been published, verify that you can navigate to it. If it was configured to appear on the Sales tab, then you should see it under this menu area—but you will likely have to scroll all the way to the right to locate it. By default it will be placed under the Extensions header (see Figure 3.27). To move your entity under a different header, see the next section.

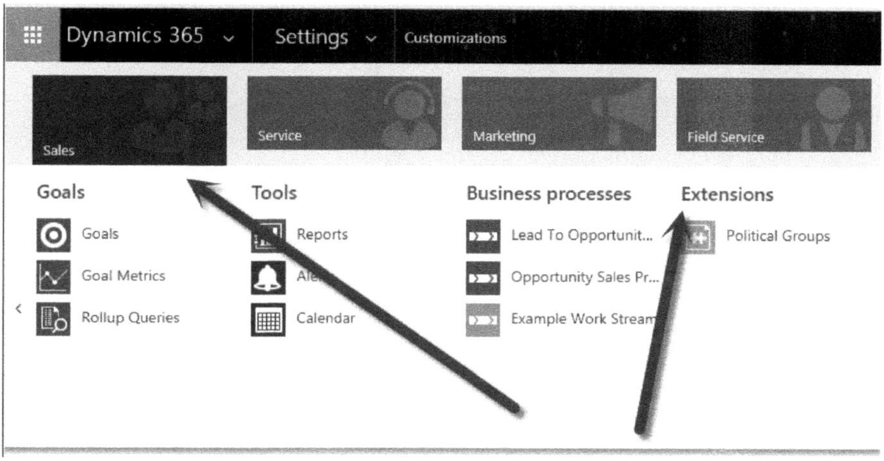

Figure 3.27: By default, your entity will appear under a subheading called Extensions.

Modifying the Navigation (Site Map Revisions)

You most likely will want to change where your new entity is located—having all of your custom entities appear under the Extensions section likely won't make sense to your users. And if you are creating new entities, it probably means that the default navigation menus are not going to accurately reflect your business as well. To modify your navigation, click on the Client Extension tab in your solution and then click on the Site Map (refer to Figure 3.28). Once selected, you can click the Edit button.

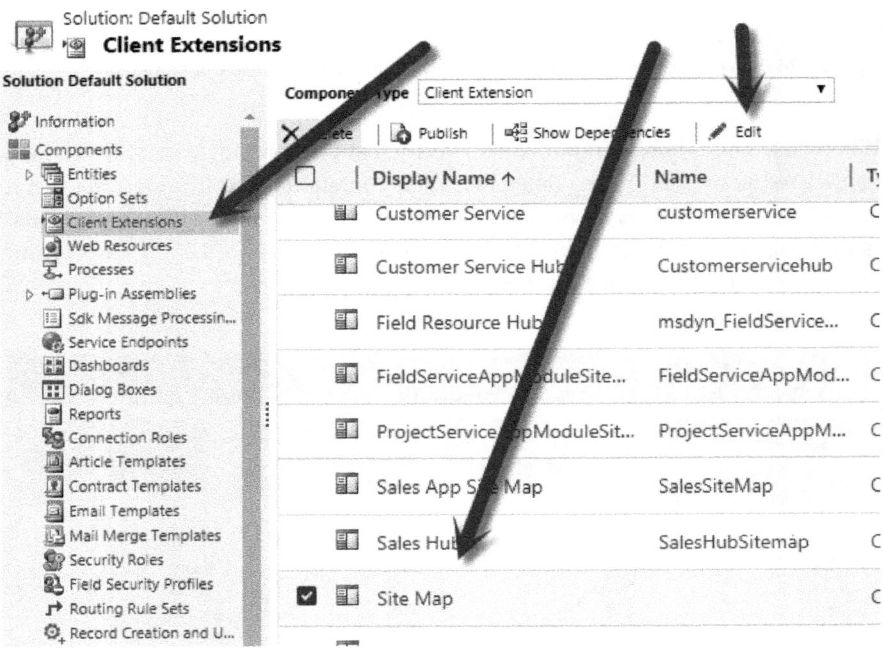

Figure 3.28: Editing the Site Map.

The screen shown in Figure 3.29 opens when the Edit button is clicked. This screen allows you to configure virtually anything in the layout and naming of the site map.

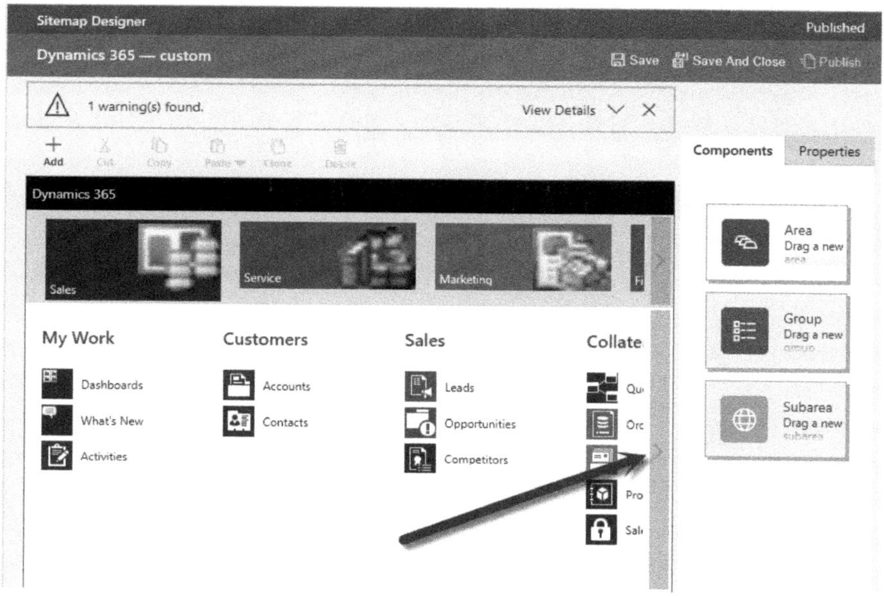

Figure 3.29: The site map editor.

As an example for making an alteration within the Site Map editor, we'll change where the new Political Groups entity is displayed. Make sure Sales is selected and then scroll to the right in the lower sub menu section. You will see it showing under the Group_Extensions header. You can either drag and drop your Political Groups tile to another location, or you can click on the header and rename it. Clicking on the header will open the Properties tab on the right hand side (see Figure 3.30) where you can set the Title and the ID.

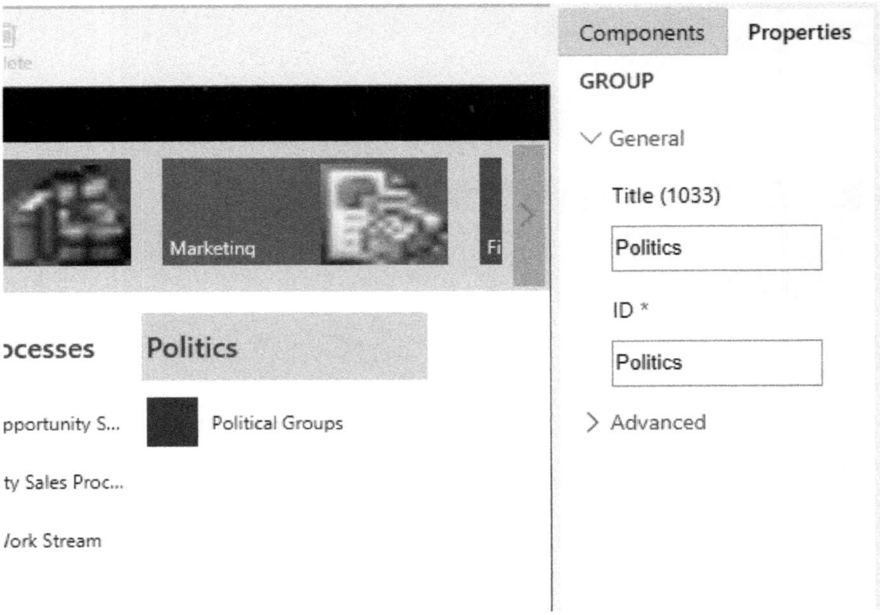

Figure 3.30: Renaming a group.

There are many things you can do within this menu. The main toolbar of the menu lets you add a new area, group, or subarea; cut, copy, paste, clone, and delete. The right-hand menu lets you add areas, groups, and subareas (just like the Add button, only drag-and-drop) and modify the properties of components on the menu.

You very likely will have areas of CRM that you are not going to use. For example, when you first set up the environment (as shown in Chapter 1), you may have opted to turn on all functionality in your new Dynamics 365 environment. If you did so, things like Field Service, Project Service, and Resource Scheduling would have been enabled. You may decide that you do not want the functionality that these areas provide access to, so you can delete them.

When you are done modifying your site map, click the Save button and then the Publish button (Figure 3.31). Once published, you will see that your modifications have taken place once you refresh your main CRM window.

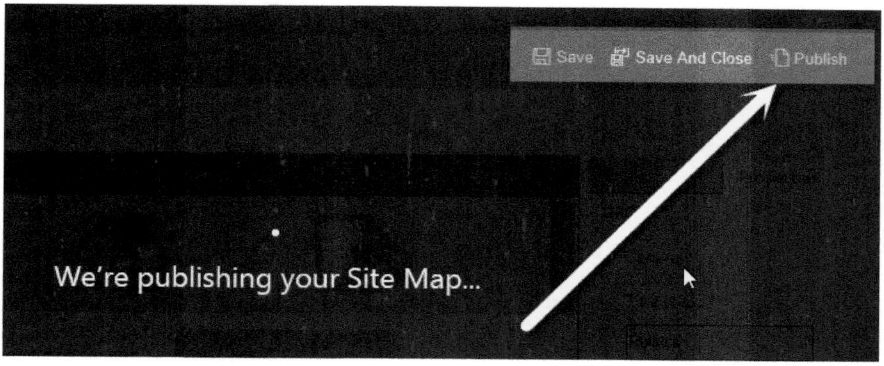

Figure 3.31: Publishing the site map.

IFrames

If you would like to display other web pages within your form, you can most easily do this by adding an *IFrame*. An IFrame can contain any web content. If you want to add a simple URL, you can open your form editor, click on a section where you want the IFrame added, and then click on the Insert tab on your ribbon and click on the IFrame button (shown in Figure 3.32).

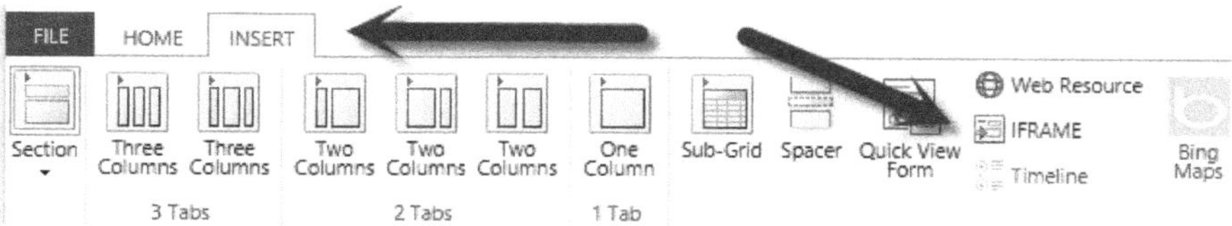

Figure 3.32: Inserting an I frame.

Once you click the IFrame button, you will be able to configure it. To configure it to show static content, simply type in the URL of your target page in the URL property (after naming the IFrame) and click OK (see Figure 3.33). For example, you might want to have a static page to link to your company portal or a specific location within that portal that has documents or links specific to the type of entity you are on.

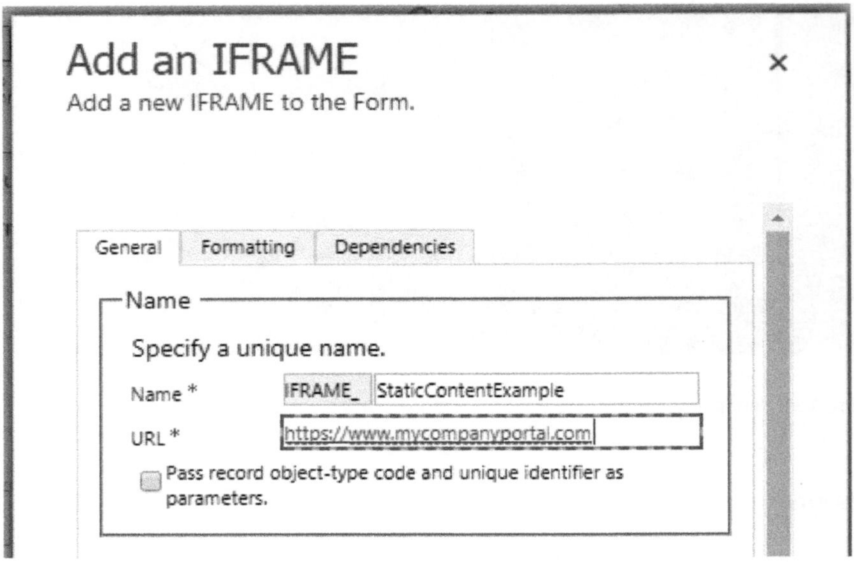

Figure 3.33: A static IFrame.

Static content within an IFrame has limited value. In most cases, you'll want to use IFrame to display dynamic content through the use of a custom web page that you have developed. For example, you might need to access shipping rates from a third-party web service when you are creating an order. On the Order entity, you could add an IFrame and point it to a custom web page that you have created. That web page is .ASPX code and is completely developed outside the scope of CRM; its function is to make a call to a vendor web service and get back shipping details, displaying those details in a grid. You would develop and test this code entirely outside of CRM. Once it is complete, you could add it to an IFrame on the Order entity.

In order to have dynamic shipping rates, you can pass in parameters from your CRM instance. By clicking the "Pass record object-type code and unique identifier as parameters" checkbox on the IFrame properties, a number of variables will be automatically appended. The URL shown below shows the OrgLCID, the UserLCID, the GUID for the record that is currently showing, the orgname, the type, and the typename. In your custom web code, you could read these parameters from the URL and take an action based on them.

```
https://www.mycompanyportal.com/?OrgLCID=1033&UserLCID=1033&
id=%7b9B16A357-C2BF-E711-A950-
000D3A328ED9%7d&orgname=org379fa4f7&type=1&typename=account
```

To see the URL and the parameters being passed, publish your IFrame and then load a record with the IFrame on it. In the IFrame, right-click the content and select View page source (see Figure 3.34). This will open up the source in a new tab, and the address bar will contain the full URL with all variables.

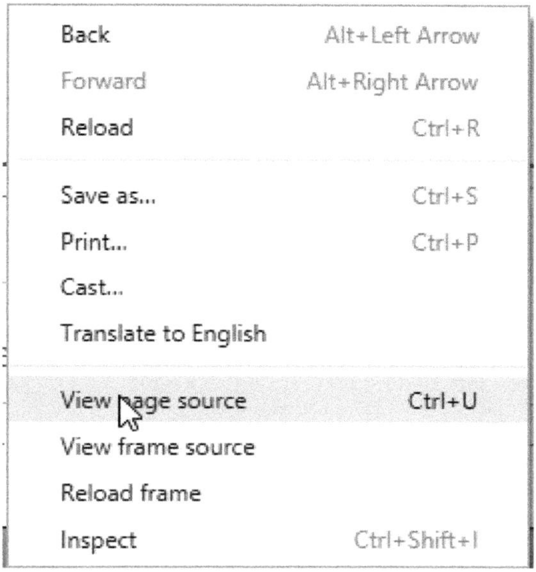

Figure 3.34: Viewing the source of the IFrame.

While some basic functionality is available through this dynamic passing of parameters, you very likely will have web code that needs additional information to proceed. In order to pass in more detail to the IFrame, you would need to use custom code in JavaScript, and tie it to the OnLoad event of the form. That way, when the form opens, it will automatically create the dynamic URL and force the IFrame to load—all through code.

JavaScript

In earlier version of Dynamics, JavaScript was heavily utilized. You will still have situations where you want to use JavaScript, but many of the original functions of it, such as hiding fields when certain conditions were met, or showing values in option sets when other fields had values selected in them are often more easily done by using business rules (see Chapter 6 for more information on business rules). That being said, if you are a developer, or you have some complex logic that must be implemented on a form, you have the ability to do virtually any client-side action you want using JavaScript.

To illustrate how to use JavaScript, we'll dynamically set the URL of the IFrame that was just created. The IFrame has a static URL which is configured when it is created, but this can be overridden through JavaScript, and it can be done at the time the form is loading.

To begin adding JavaScript to your form, open your editable form. On the toolbar, click the Form Properties button (see Figure 3.35).

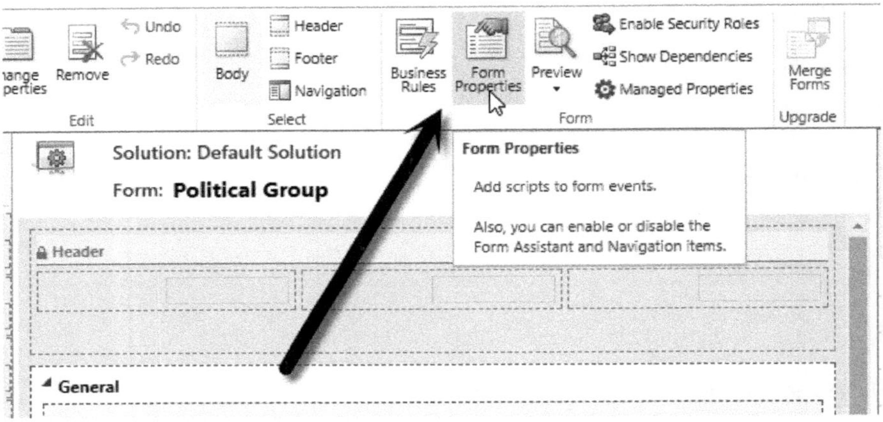

Figure 3.35: Accessing the Form Properties.

In the window that opens, you will default to the Events tab. On this tab you will be able to add a new *library* to your form. In this case, a library will be a JavaScript file, that will be stored as a web resource within Dynamics. When you have the library defined and the code added to it, you will then create a new event handler to indicate what code to execute and when. A library can contain many functions. In this example, we'll create one function and trigger it when the form loads. To create a new library, click the Add button (Figure 3.36).

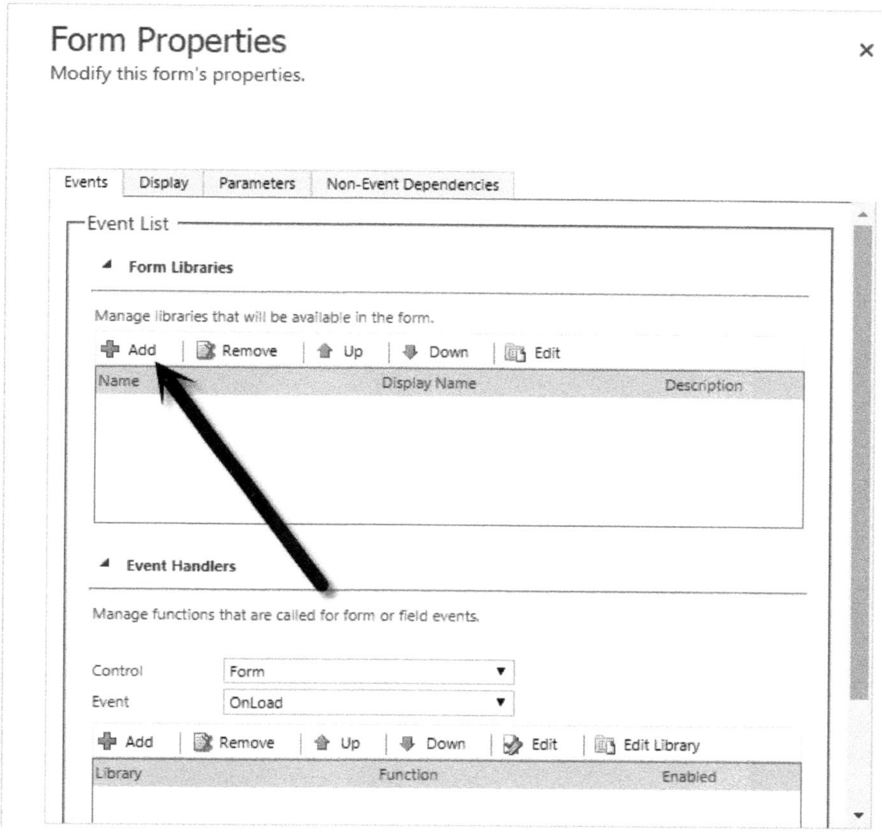

Figure 3.36: Adding a new library.

Clicking the Add button will open a lookup window where you can add a reference to an existing web resource or create a new one. We want to create a new one, so click the New button in the lower left hand corner (see Figure 3.37).

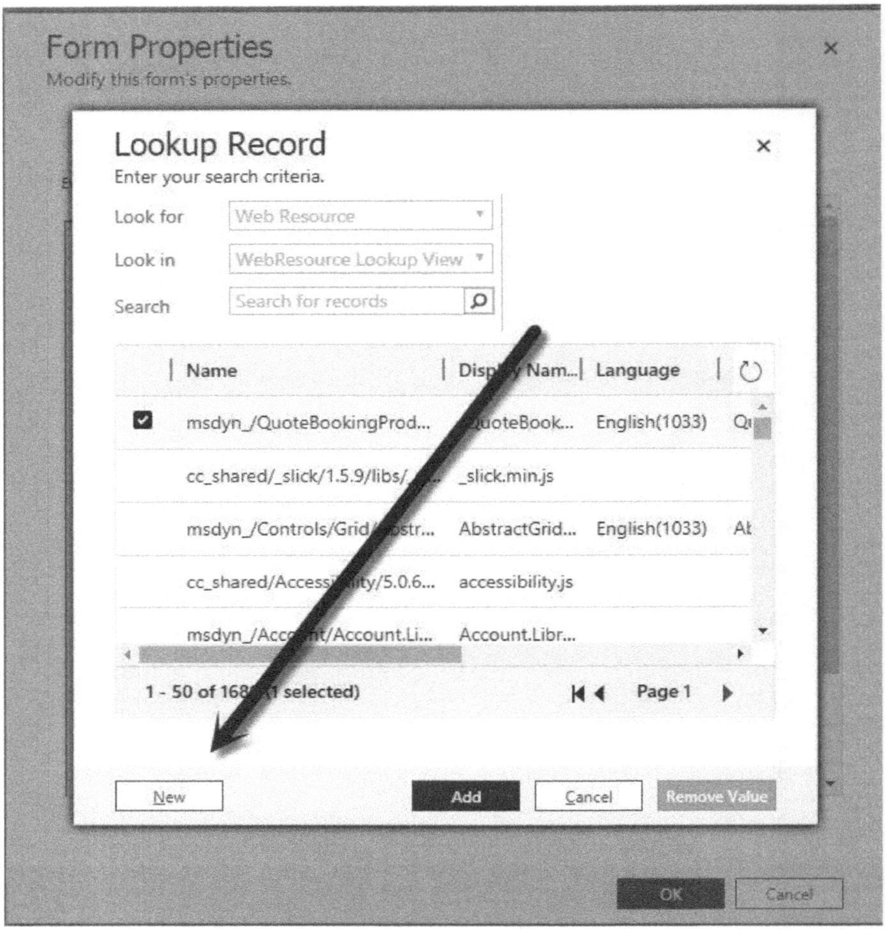

Figure 3.37: Click the New button to add a new web resource where you can code your JavaScript.

Another window will open (Figure 3.38), where you will be able to create your web resource. This will contain the JavaScript that you'll reference from your form. You will create a file that you can use across all your forms, and will allow you to write a function once and call it from a variety of locations. In this case, you can name your Web Resource whatever you would like, and select Script (JScript) as the Type.

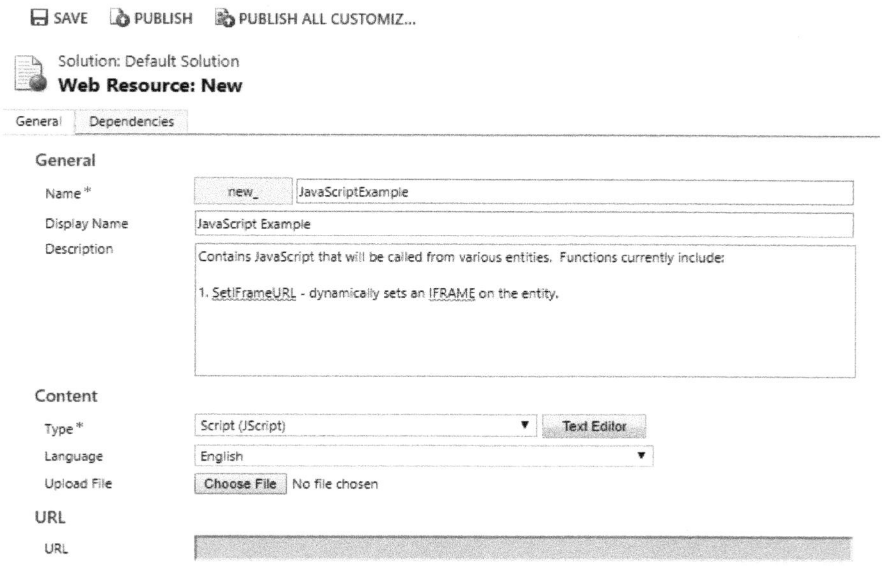

Figure 3.38: Creating a script file for your JavaScript.

You can now click on the Text Editor button next to the Type property that you just set. This will open a screen where you can type your code. Listing 3.1 shows a simple JavaScript function that will set the IFrame's URL. Copy and paste this code into the content editor, as shown in Figure 3.39.

Listing 3.1. Simple JavaScript to Set an IFrame

```
function SetIFrameUrl() {
 // you could pull the IFRAME that you want to set by name
(as shown here)
 // or through logic
 var frame =
Xrm.Page.ui.controls.get("IFRAME_StaticContentExample");
 // the following sets the source of the IFrame. You could
base this
 // on some logic, or you could create a dynamic value in
the URL
 frame.setSrc("www.mynewurl.com");
}
```

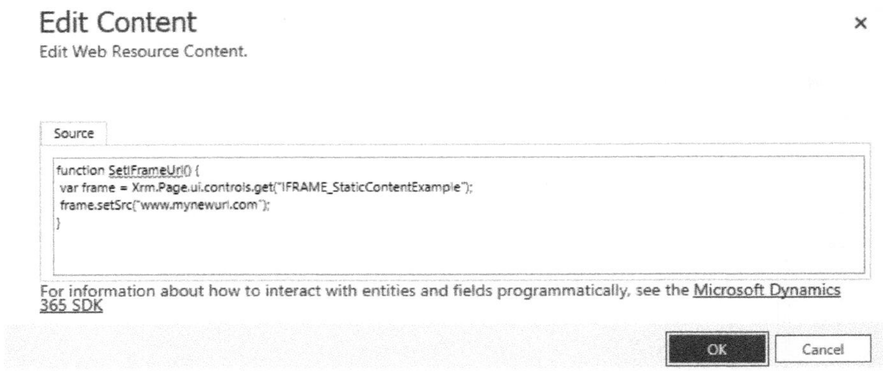

Figure 3.39: Adding your code to the editable window.

Once the script has been entered, click OK and return to the Web Resource page. On this page, you will now see a URL which means your script was saved. You can click on the Save button and then on the Publish button on this window (Figure 3.40).

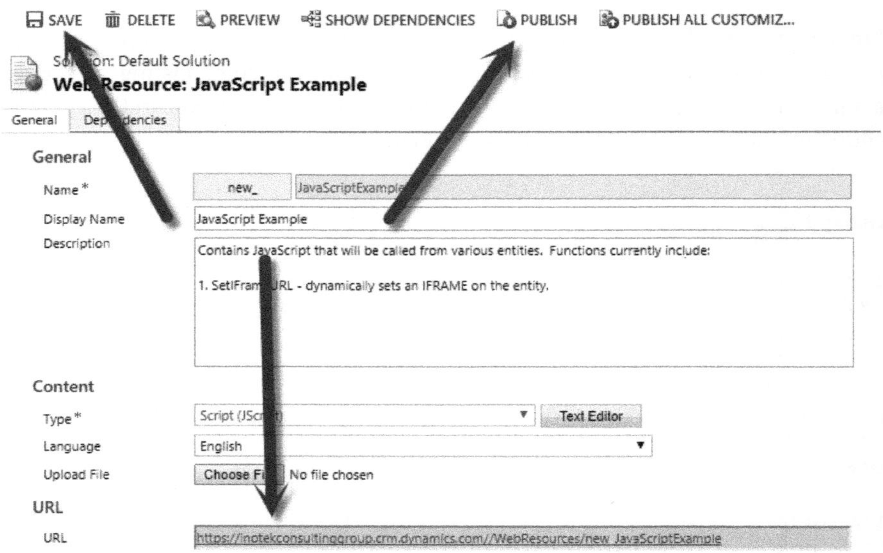

Figure 3.40: Saving and publishing your code.

Once published, go ahead and close this screen. You will now see that this new web resource you just created is selected in the Lookup Record window. Click the Add button (see Figure 3.41).

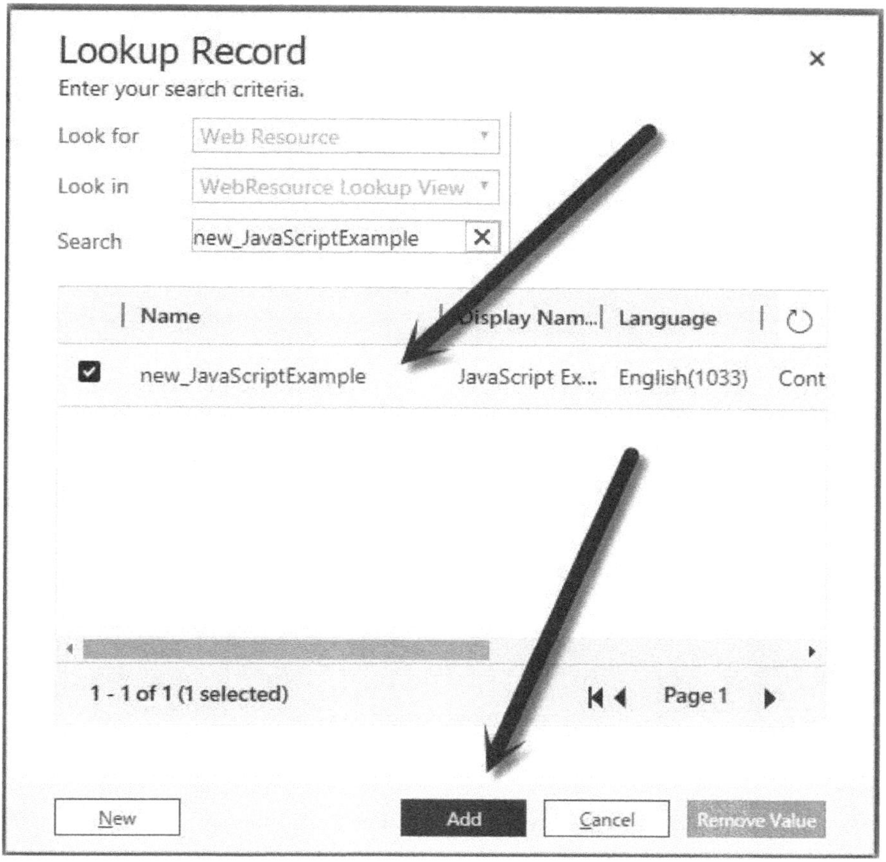

Figure 3.41: Selecting the new web resource you just created.

You should now be back on the Form Properties screen, which will have the web resource you just created added to the Form Libraries section. This will allow you to configure the Event Handlers section of your form. You want your code to execute at the time your form loads, so set the Event property to OnLoad and then click the Add button (see Figure 3.42).

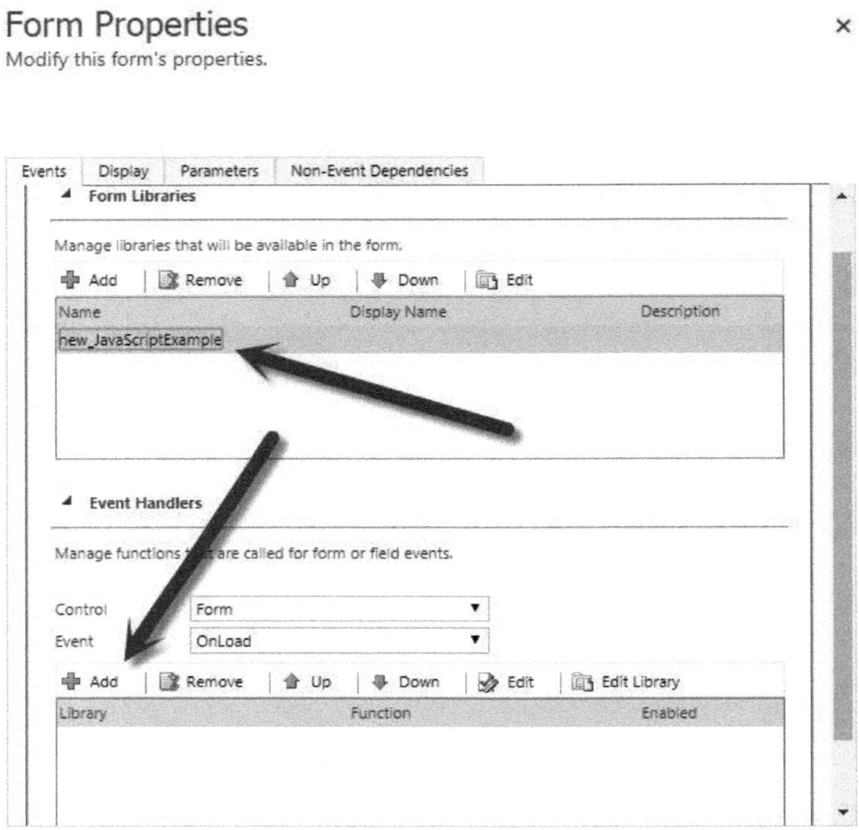

Figure 3.42: Now you have to tie a handler to the code you wrote.

When you click Add, a window called Handler Properties will open. This will show your library selected (others could be selected if you had anything else added). You will then be required to type in the name of the function you want to call when this event triggers. The code that you entered above in Listing 3.1 has a function called SetIFrameUrl, so this would be the value entered for this example. Click the OK button on the screen shown in Figure 3.43 when you are done.

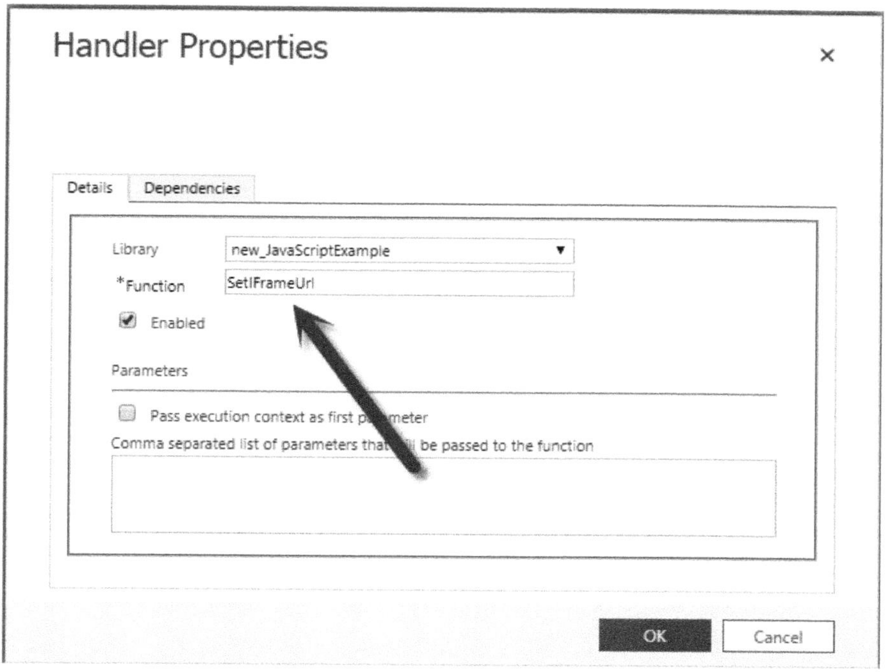

Figure 3.43: Calling the function you just created.

Your Form Properties screen (Figure 3.44) should now be fully configured. You have added a new library with a single JavaScript function in it, and you've told Dynamics to execute this function when the form loads.

If you want to add more JavaScript functions, you do not need to create a new library next time. You can simply open the library that you already created and add your new function to it. You can also reference this JavaScript web resource from any entity.

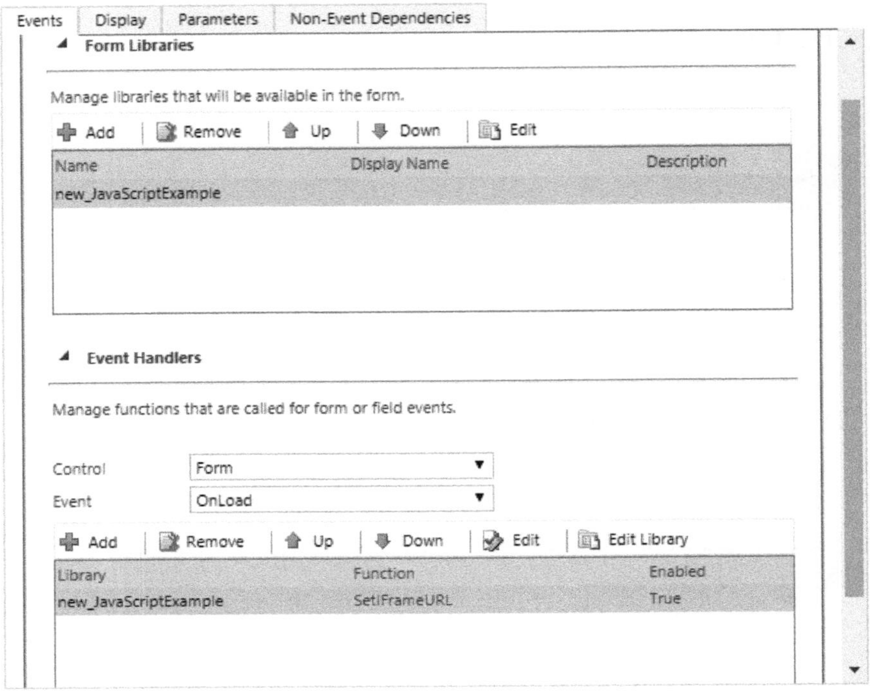

Figure 3.44: The form is now fully configured to call the JavaScript on load of the form.

On your main form, you can now click Save and then Publish. You should now be able to open up a record and have your IFrame automatically populated based on the URL you placed in your JavaScript.

If you get a Script Error (like that shown in Figure 3.45), you most likely have a syntax error in the JavaScript. Check that you have typed in your function name properly, and that you don't have any errors in the code you have written. You may also need to publish all your customizations, as the web resource file may not have the latest code published. In the case of the error shown below, the parameter was set to a function that doesn't exist. It shows "SetIFrameURL" (capital URL letters), whereas the function name is actually "SetIFrameUrl". In order to resolve an error like this, the name would need to be updated, and the change would need to be republished.

Figure 3.45: If you get an error, click the Download Log File for more details.

Creating a Personal View

Views are everywhere in CRM. When you click on a menu item in the main navigation, you are presented with a view into your data. You can see an example of these views when you open your Accounts from the Sales menu, shown in Figure 3.46. As you can see, there are many views here. All of these views are "out-of-the-box" views, which means they present information as they were configured by Dynamics 365 when you set it up. You have the option to modify these views, or create your own views. Views can be made private (where only the user that created it can see it) or public (share with one or more people). To create a new Personal View, click on Create Personal View in the list of views. This option appears at the bottom of the list (refer to Figure 3.46 again).

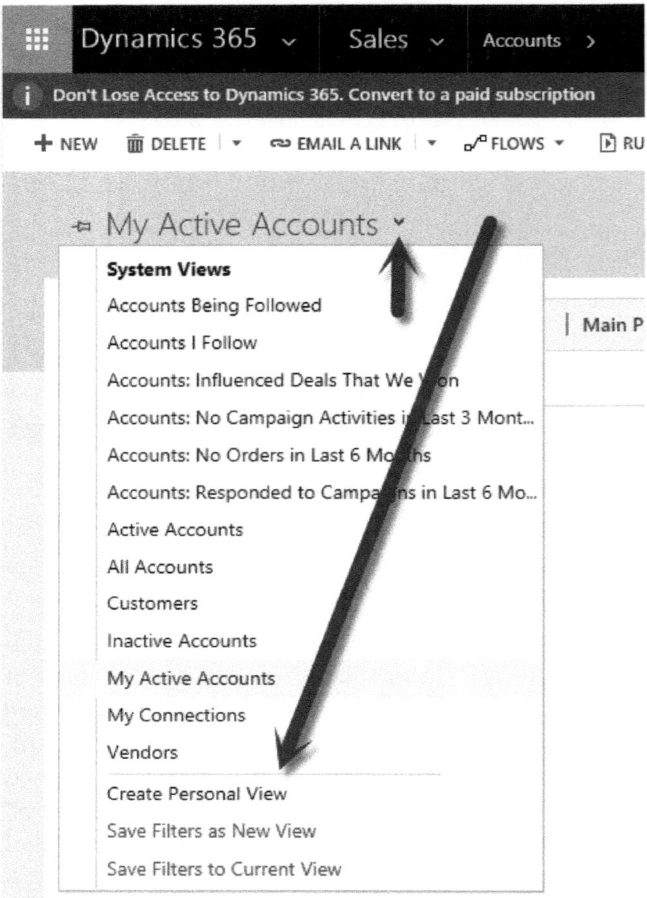

Figure 3.46: Creating a personal view.

Clicking Create Personal View will open an Advanced Find window. You can define whatever information you want in your advanced find query, and then you can click the Save button. You will then be able to name the view, like that shown in Figure 3.47. When you have set the query and named it, you can click the Save button.

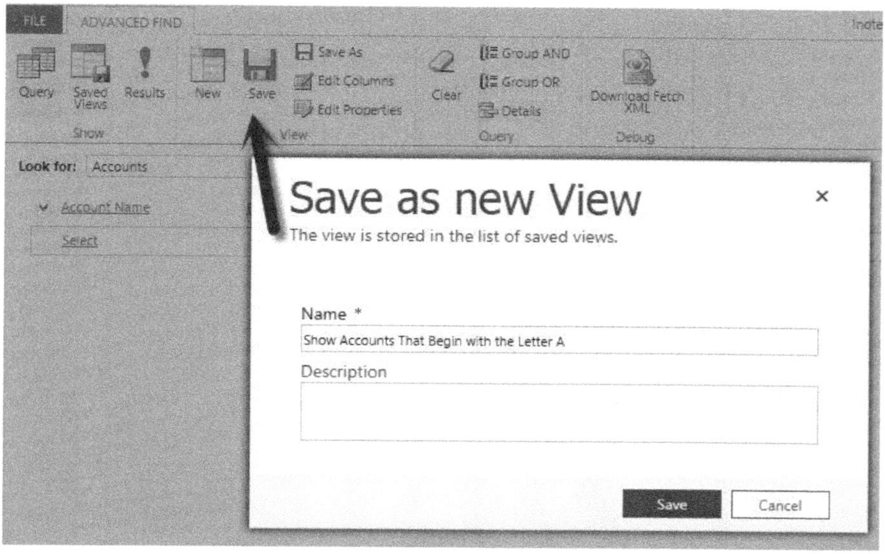

Figure 3.47: Provide a name—only the user creating the view will be able to see it.

After you refresh your main Dynamics browser window, you will see this new view available in your drop-down list (Figure 3.48) under the My Views heading.

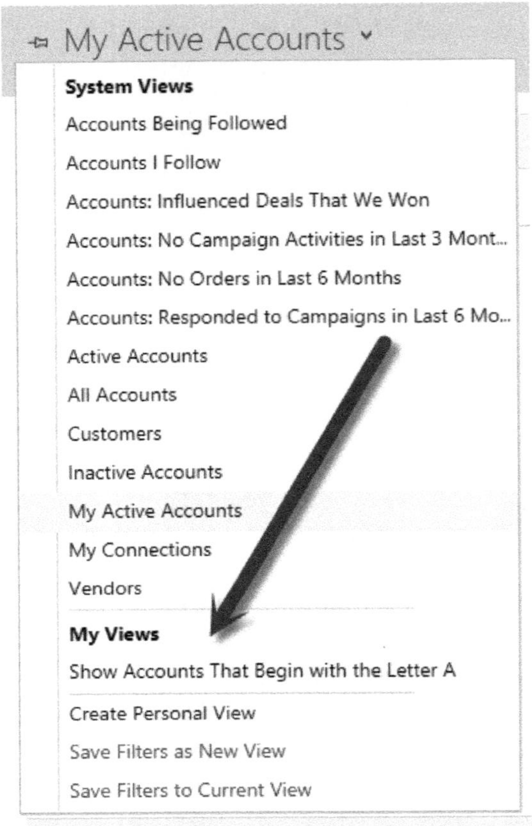

Figure 3.48: The newly created personal view.

Customizing a View

Being able to create a personal view is useful to an individual, but what about if you want to make a change to an existing view that all users can access? For example, let's say you want to add an additional column to the "All Accounts" view. By default, the view shows the columns in Figure 3.49. Now you want to modify it so that the second column showing is a new value not currently on the view. Here are the steps you can take to add this new column—for this example, we'll add the Annual Revenue field.

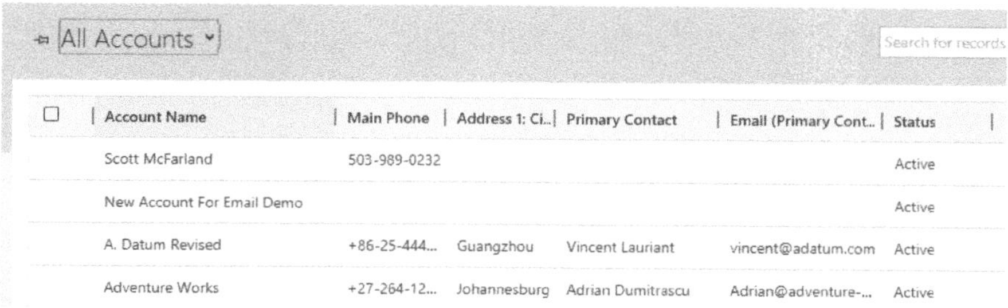

Figure 3.49: The existing All Account view—we'll add a new field.

1. In your solution, expand the Account entity, and click on Views. In the list, select the check box next to All Accounts and then click More Actions, followed by Edit (see Figure 3.50).

Only views that show True under the Customizable column can be altered.

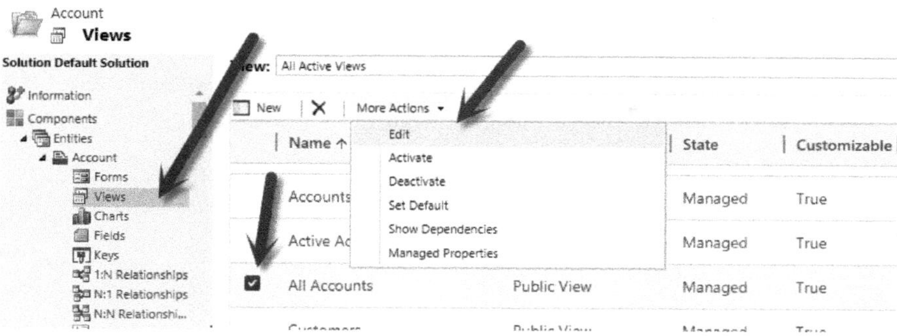

Figure 3.50: Selecting the view to edit.

2. In the window that opens, click on the Add Columns button in the Common Tasks menu on the right-hand side of the screen (see Figure 3.51).

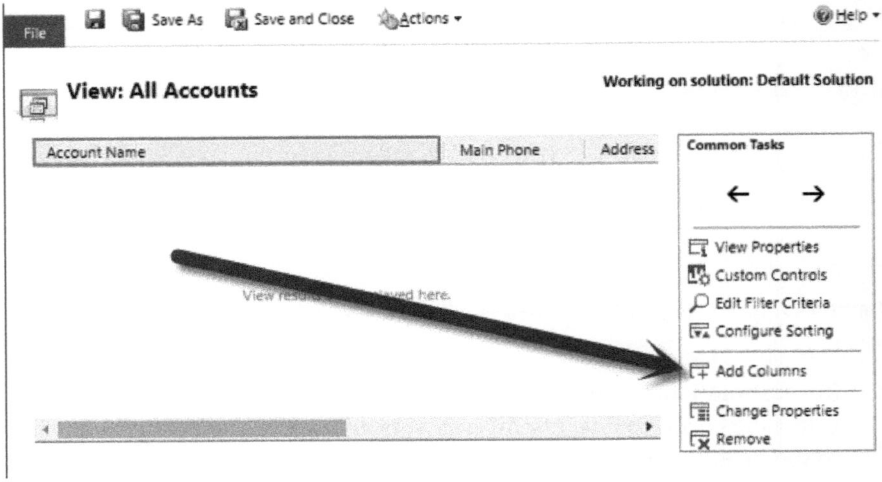

Figure 3.51: Adding a column to the view.

3. There is a small screen that opens titled Add Columns, and you can scroll through the available fields to select the fields you want to add to your view. In some cases, depending on the entity, you can also add columns from associated entities into the view (see Figure 3.52). For this example, select the Annual Revenue field and click OK.

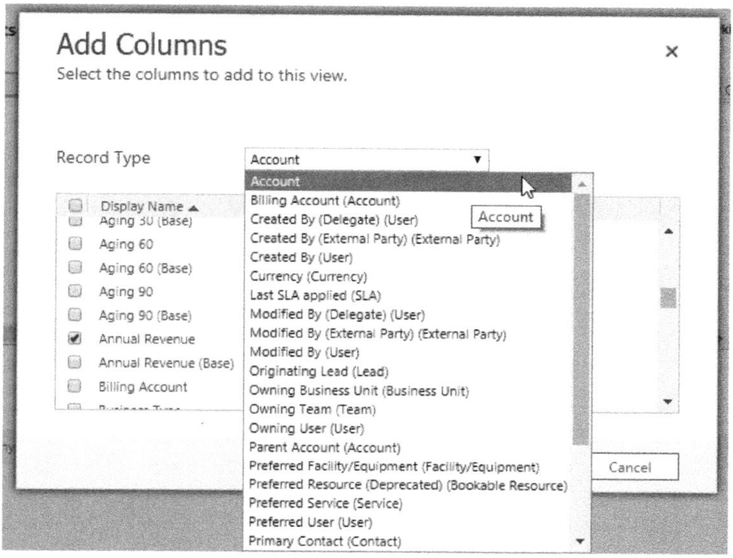

Figure 3.52: Selecting the Annual Revenue field from the account.

4. The field will now appear as the second column in the view, assuming you had clicked on the Account Name before you added the field. If not, you can use the arrow keys under Common Tasks to move your column around. Figure 3.53 shows the Annual Revenue in the second column. Click Save and Close on this screen when you are done adding your field.

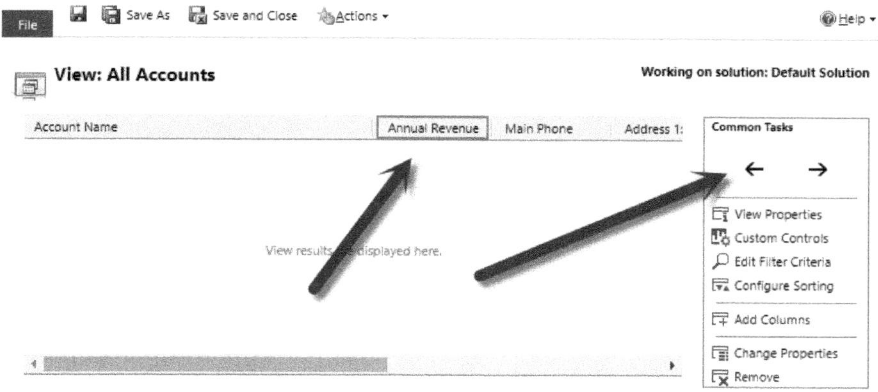

Figure 3.53: Moving the Annual Revenue field to the second column with the left and right arrows.

Back on the main solution, you need to publish your customizations. Once published, your view will have the updated column showing in it, as seen in Figure 3.54. You have to refresh your browser to see this modification.

Figure 3.54: The new field on the original view.

Creating an Editable View

To create your own new view, take steps similar to how you modified the existing view. Go in to your solution, click on views, and click the Add button. A View Properties screen will open where you can enter the name and description of your view (Figure 3.55).

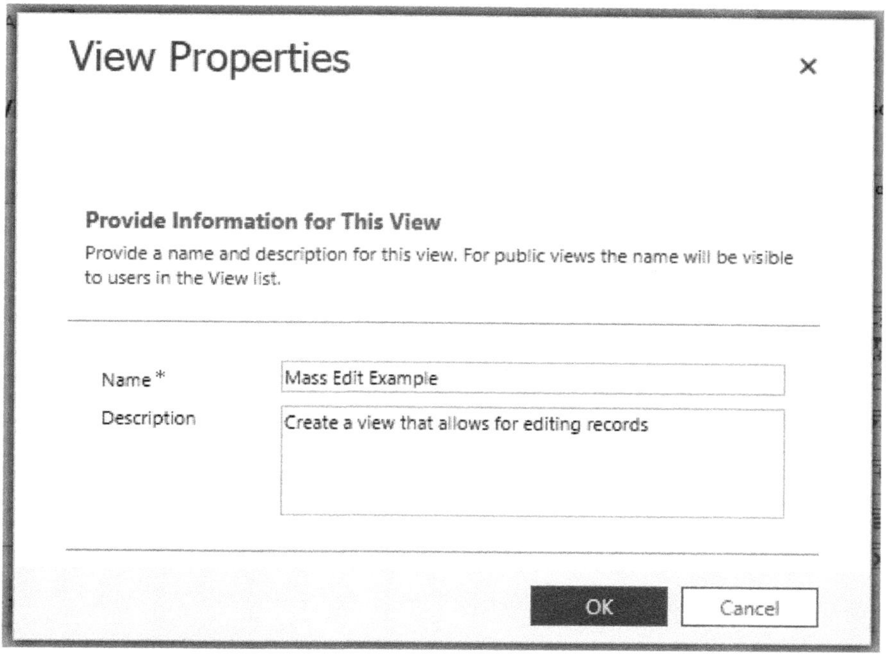

Figure 3.55: Creating a new editable view.

When you click OK, you will be in the view configuration screen that you are already familiar with from the earlier example in this chapter. However, this time we are going to create a view that allows for an editable grid. Click the Custom Controls button in the Common Tasks, and then select Add Control on the Custom Controls window that opens. In the Add Control window, you will be able to add an editable grid to the view, as shown in Figure 3.56. This figure also shows some of the other controls that can be added to a view.

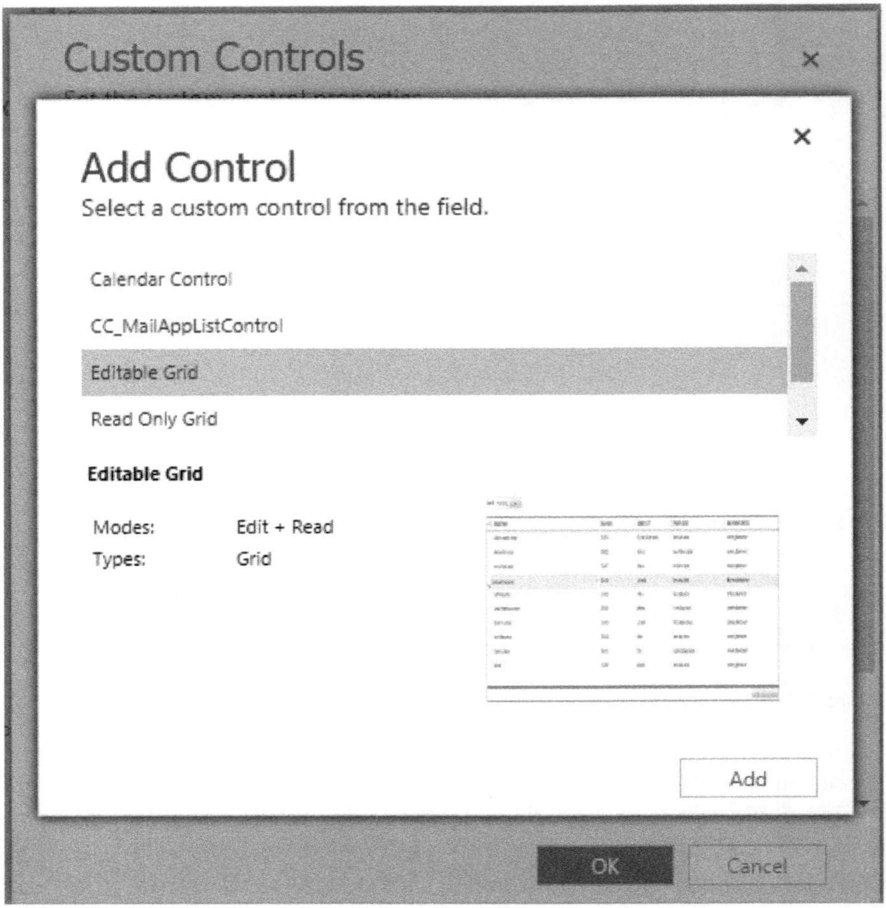

Figure 3.56: Adding the Editable Grid control to the view.

Once you add this editable grid, you will have some ability to alter how it works. You'll be back on the Custom Controls page (Figure 3.57) where you first want to select all the device types where this control will be allowed. In this example, we'll select the Web.

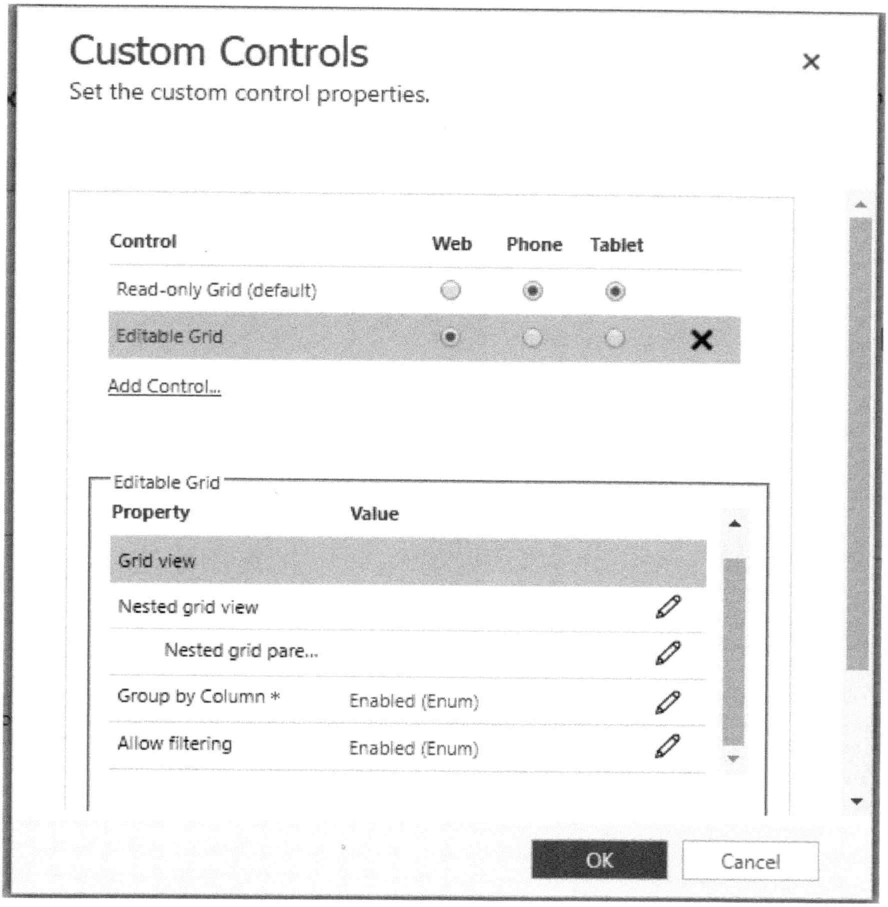

Figure 3.57: Customizable properties on the editable grid control.

Click OK and return to the main view configuration screen. Two additional things that you will want to do in your new view include:

1. Add additional fields that you'll want to be able to edit. In this case, we'll just add the phone number as a new column.
2. Click the Edit Filter Criteria to limit the accounts that are returned. In this case, we'll only return accounts that have no address data. See Figure 3.58.

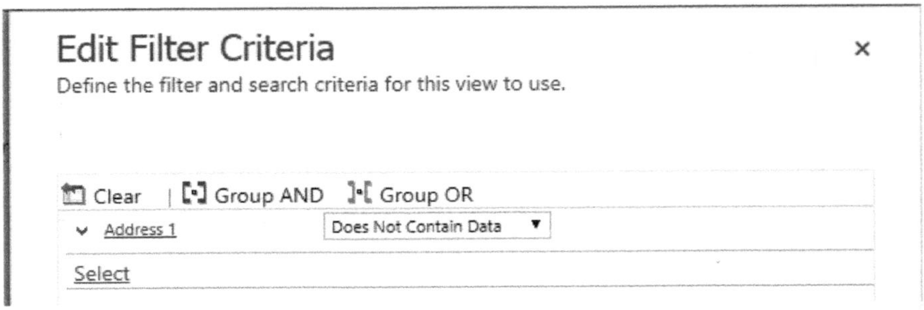

Figure 3.58: Limiting the search results.

Now click the Save and Close button and publish your view.

Next, you will have to activate it. On the list of views in the solution, click next to your new view and click on More Actions and select Activate (as shown in Figure 3.59).

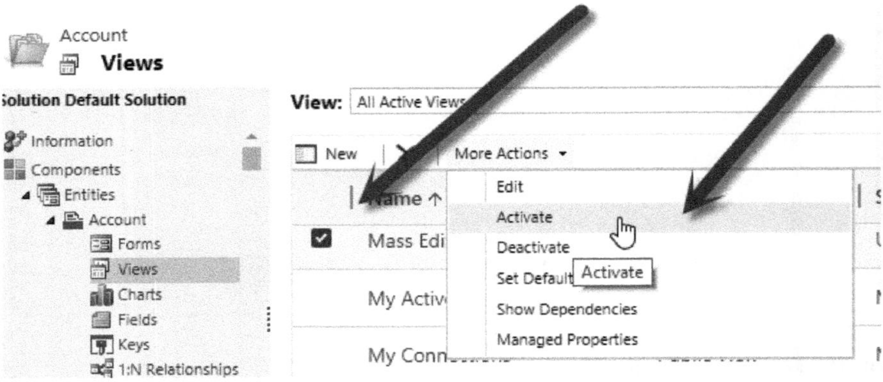

Figure 3.59: Activating the view.

You will also need to go to the Account entity itself and enable editable grids. This means that any view that has the editable grid control added to it will allow users to edit data within it. To do this, click on the Account, and then select the Controls tab. On this tab, click Add Control... and select the editable grid, just like you did on the individual view. Figure 3.60 shows where this is set on the Account entity.

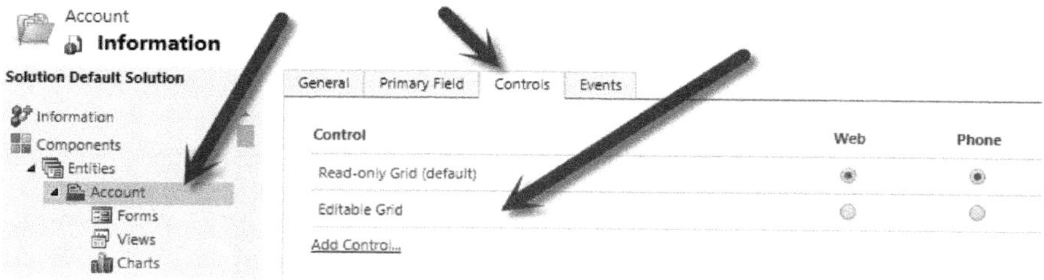

Figure 3.60: Enabling editable grids on the account entity—this must be done in addition to adding it to a view.

Publish your customizations, refresh your browser, and return to your account view list. Select the view you just created. On the main tool bar, click on the Show As button and select the Editable Grid option (see Figure 3.61). You will be able to edit data right in your view! This can greatly improve efficiencies for users who have many records that they need to work through, but don't want to click through each individual form and record (navigating around CRM can be very time consuming!). Any view with the editable grid control added to it will have that functionality.

The Show As button on the toolbar will only display after you have enabled editable grids on the entity and published that customization.

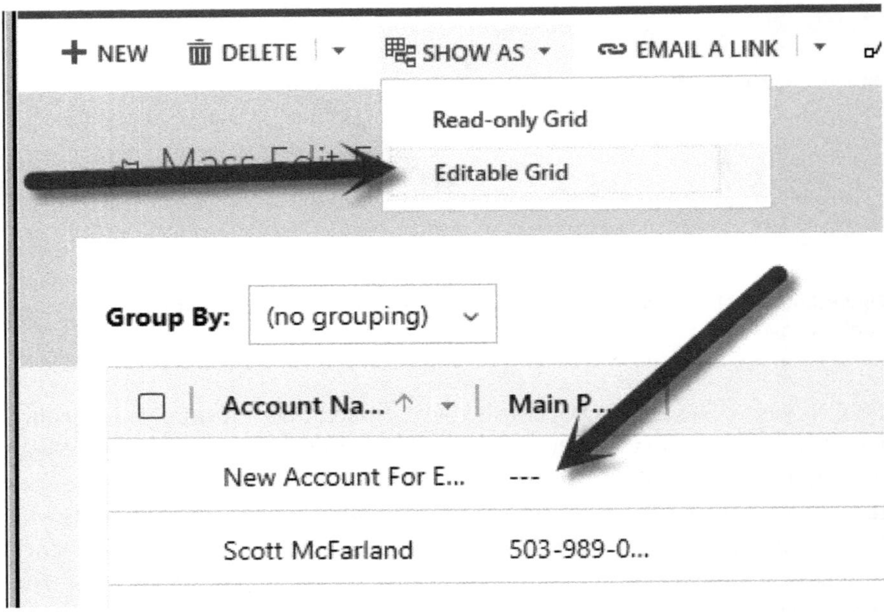

Figure 3.61: Select Editable Grid in order to allow inline editing of data.

Summary

We've covered a lot of information in this chapter. The base customizations presented are around modifying forms, adding fields, creating new entities, modifying the site map, adding JavaScript, customizing views, and working with editable grids. But, as you will see in the coming chapters, virtually everything you do in CRM is a customization of some sort. Reports, workflows, business rules, plugins—all of these are components and layers of customization of Dynamics 365. I hope you will realize that it is very easy to customize and publish your changes and that many of the changes do not require a true code developer to make them.

Chapter 4
Reporting

One of the primary differences between Dynamics 365 Online vs. On-Premises is the way reporting is done. In an on-premises environment, you have access to the underlying SQL Server database and can write traditional SQL-based queries. In online environments, no database is available, so the data either must be replicated to a local SQL environment, or it must be queried using FetchXML. This chapter will look at how to access reports within CRM and how to do basic development using the methods available. You will have two basic methods:

1. Build reports using the Report Wizard from within CRM
2. Build reports using FetchXML with a tool called the Business Intelligence Development Studio

There are several ways to access reports in Dynamics 365. The first way is to browse via the toolbar menu. Generally, you will find it by clicking on Sales and then looking under the Tools submenu (you may have to scroll to the right—see Figure 4.1). Clicking this will show all of the reports currently available in the system.

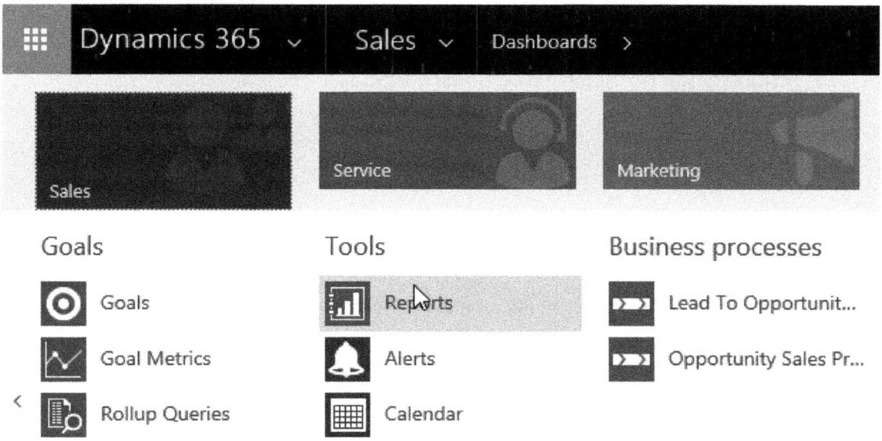

Figure 4.1: Accessing reports via the toolbar.

DOI 10.1515/9781547400065-004

You can also access reports through the Advanced Find menu, which is the quickest way to get to things. Both windows will allow you to select a report and edit it (by clicking to the left of the report name and then clicking Edit) or add a new report (by clicking the New button on the toolbar). Figure 4.2 shows the New and Edit buttons within the Advanced Find result view.

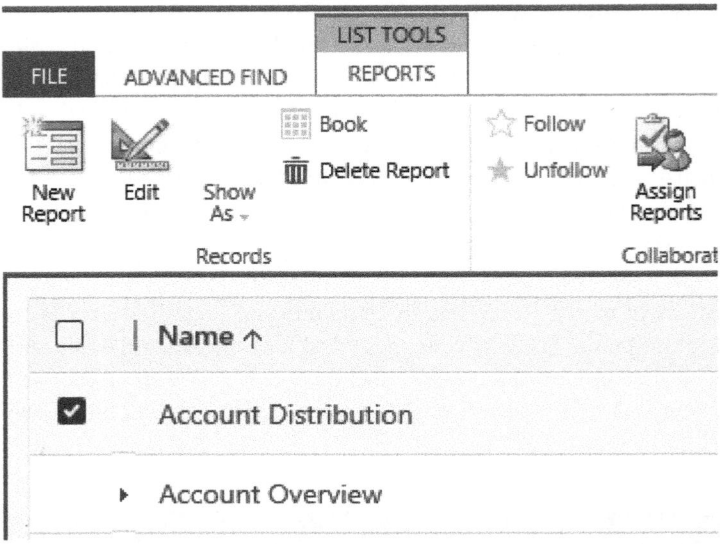

Figure 4.2: The New and Edit buttons from within the Advanced Find view.

Using the Report Wizard

There are two options for creating reports in Dynamics. The first option is the use of the Report Wizard within CRM. This gives a non-developer access to creating a report to query data that has the same complexity of logic as what can be done with an Advanced Find. The difference is that you can control the layout and appearance of the report, whereas Advanced Find will just show a list view. If you have more involved reporting needs, you will want to look at the second option covered later in this chapter, which is using the *Business Intelligence Development Studio* to create reports with more complex logic and layout.

Figure 4.3 shows the window that is opened after you click the New button from one of the report areas mentioned in the previous section. The first drop down in the window is the Report Type. Select the Report Wizard Report option and then click the Report Wizard button.

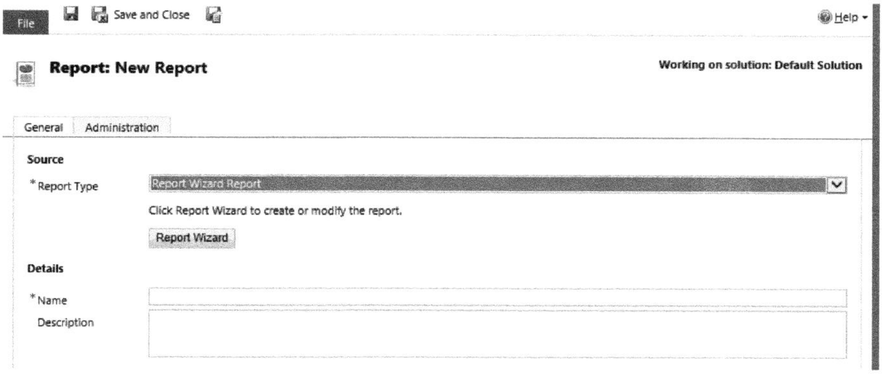

Figure 4.3: Selecting the Report Wizard option for development.

Once the Report Wizard button has been clicked, a new box opens (Figure 4.4) where you can click on Start a new report. Click Next on this window.

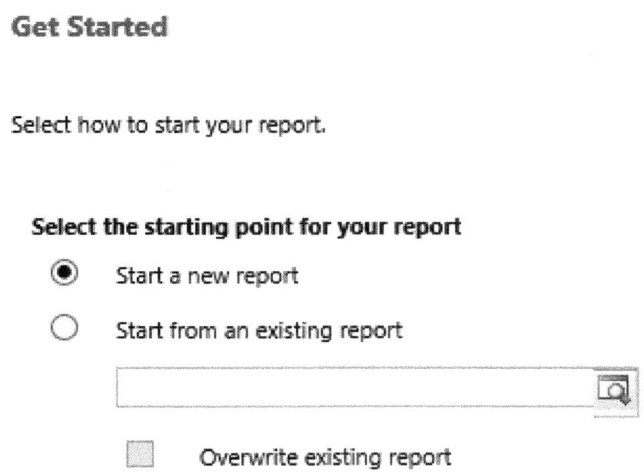

Figure 4.4: Starting a new report.

The next screen of the wizard is called Report Properties. Here you can name your report and select which entities the report will be built against. For illustration, we will show all accounts and their related contacts. Figure 4.5 shows the Primary record type property has been set to Accounts, while the Related record type shows available options. Once you have set a name and selected the appropriate entities, click on the Next button.

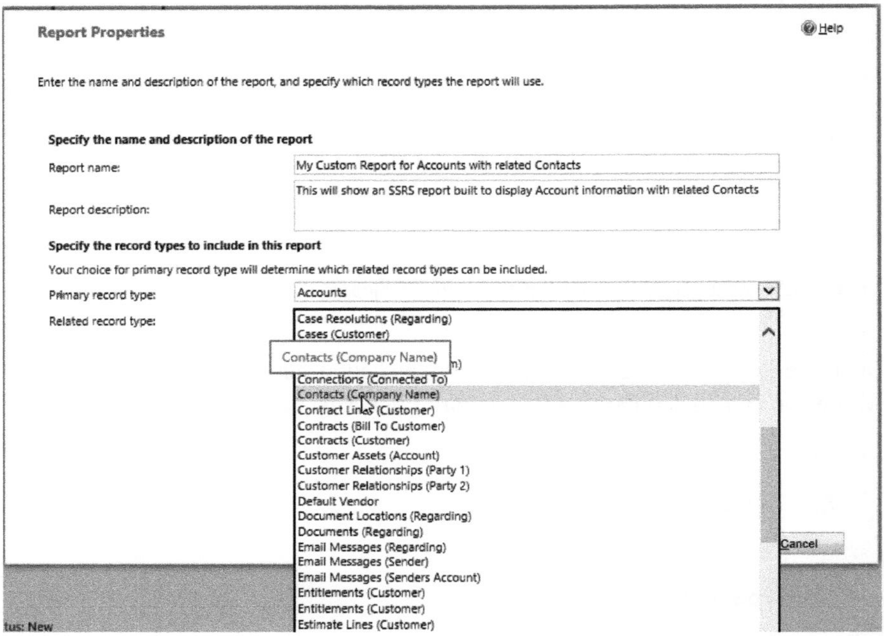

Figure 4.5: Selecting the name and entities for the report.

When the next window opens, you will see that you are given a filtering screen that looks very similar to an Advanced Find. In this case, there are two sections—the first is where you can filter on the Accounts (the primary entity) and the second is where you can filter the related Contacts (the secondary entity). Figure 4.6 shows these filters set so that all Accounts that have a value in Address Line 1 will be displayed, and all related Contacts will be returned.

Click Next once you have set the filters.

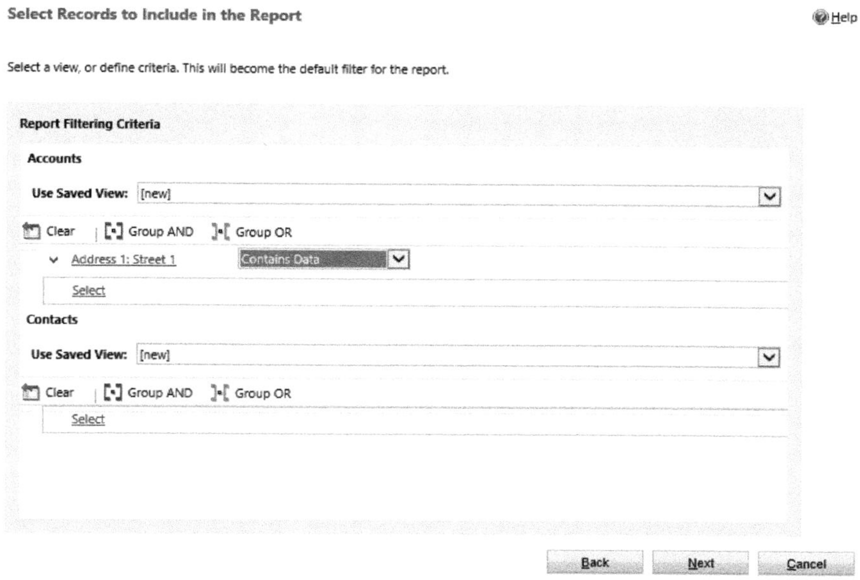

Figure 4.6: Defining the default filters.

The Lay Out Fields window will take some experimentation. You can set several levels of hierarchy, but your ability to really control the look and feel is very limited. In the example we are working with, you will want to group by the account and then show an appropriate mix of Account and Contact information. Figure 4.7 shows the available options when you click on the Click here to add a Grouping box.

In this case, we will group by an Account, with the Account Name being the data displayed. Once you have set the Grouping, click OK.

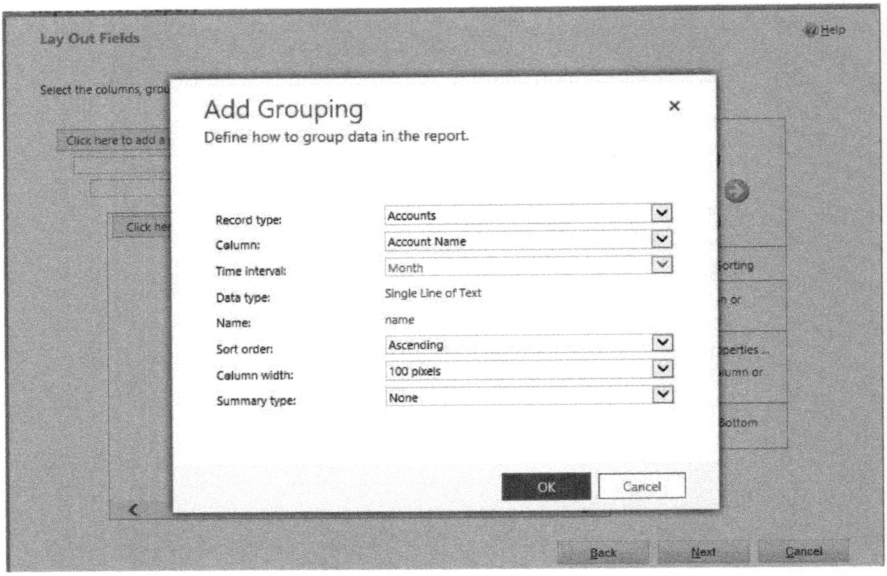

Figure 4.7: Adding grouping at the account level to the report.

You can add a second level of grouping, or you can start adding in the columns of the report. There are several options available to control your report. They are:

1. Arrows—These let you move the selected box in the left-hand window around. You will most typically use the left and right buttons to change where a column is displayed.
2. Configure Sorting—This will allow you to determine how the data is ordered.
3. Add Column or Grouping—You can add as many columns as you would like. There are several layers of groupings that you can create, but you'll most likely want to limit yourself to one or two levels.
4. Change Properties—This will let you change some limited properties, such as how wide a specific column will be.
5. Remove Column or Grouping—You can delete information that you have added.
6. Set Top or Bottom Number—You can decide if you want only a limited number of records. For example, if you want the first 10 records or last 10 records returned from the report. This has very limited value, but if you are returning thousands of records, you may want to limit it to the first 500. A better approach is to add additional filters to the query if you want fewer records returned.

Figure 4.8 shows one potential layout of the Account/Contact report. The grouping is on Account, while the column list is the name and address information for the contacts related to each account.

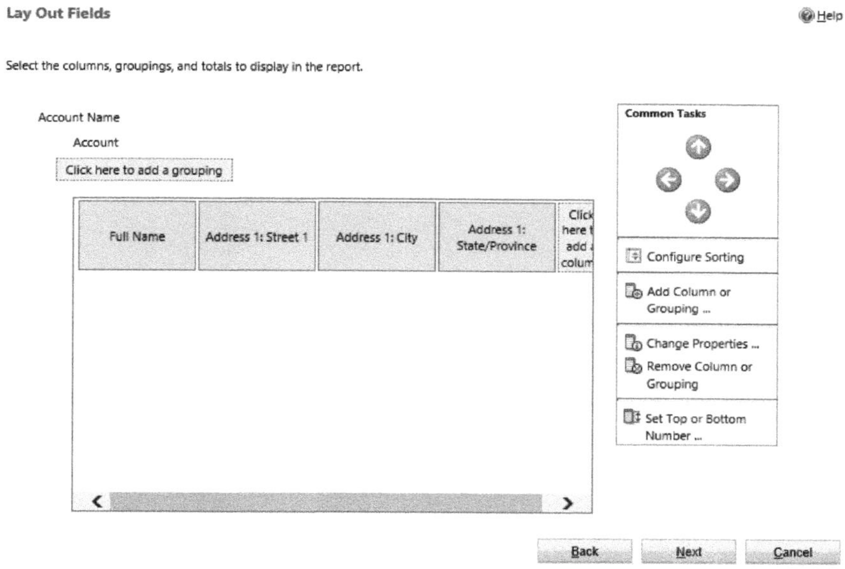

Figure 4.8: The Account is the header; the Contact is the detail at the column level.

Clicking Next on the Lay Out Fields form will take you to the Format Report form. You can select whether the report will be a table grid, a chart, or a combination of both. For the example of Accounts and Contacts, you'll only be able to select the table view, as shown in Figure 4.9.

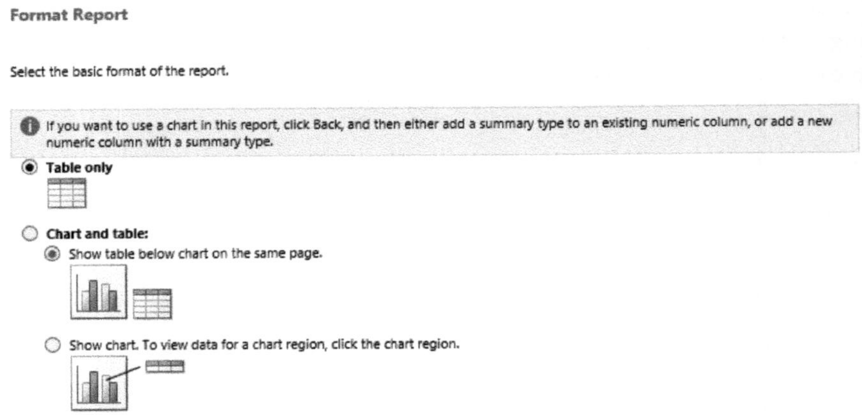

Figure 4.9: You can have a table or chart view for your report.

Clicking Next from the Format Report form will take you to the Report Summary page (Figure 4.10).

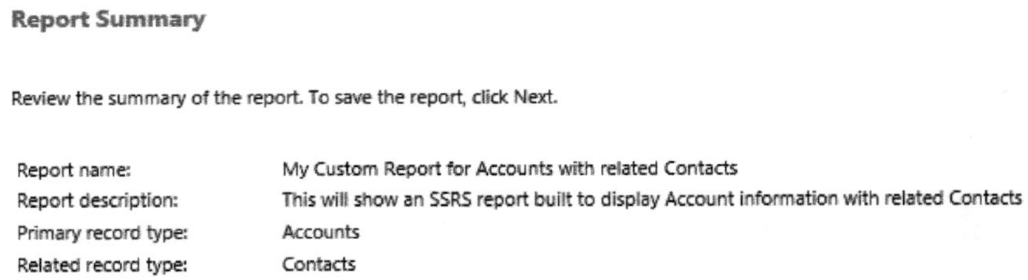

Figure 4.10: There isn't much to creating a report with the report wizard—this is the summary view.

On the screen that follows this Summary, you can click Finish. Your custom report will have been created, and will now appear as an available report when you are looking at your report list. The final configuration is shown in Figure 4.11 and the report run from CRM is shown in Figure 4.12.

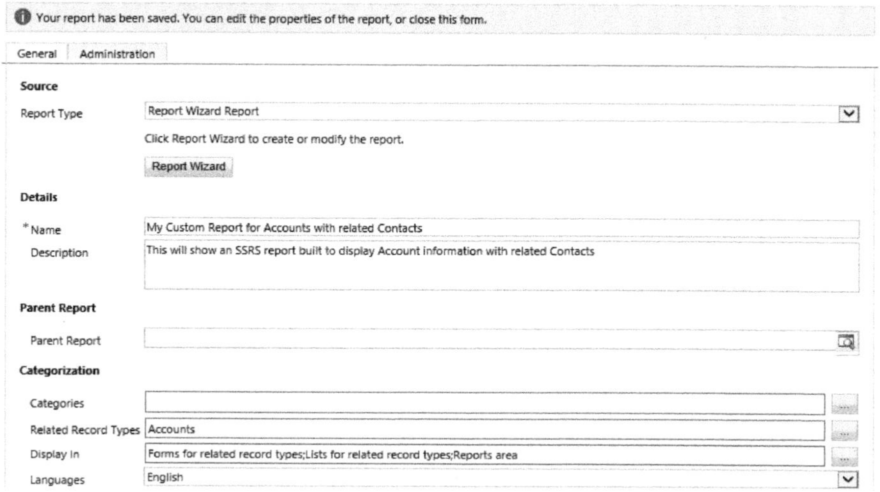

Figure 4.11: The configurations of the finalized report.

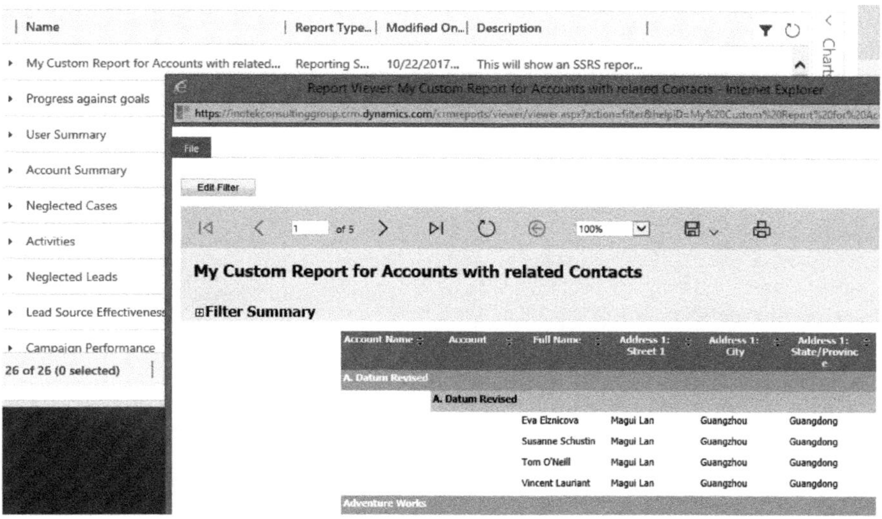

Figure 4.12: The report is available from the report list, and looks like this when run.

As you can see, there is not a whole lot of control you can have over a report using the Report Wizard approach, but it does give a non-developer some ability to create a report. For most reporting needs, you will want to move on to a more advanced development platform, which is covered in the next section.

Using Business Intelligence Development Studio (BIDS)

Building reports that will be accessible from within CRM can also be built using Microsoft Visual Studio and the Business Intelligence Development Studio (BIDS) platform. This tool allows you to develop and test reports that run against either Online or On-Premises solutions. When you are developing against a Dynamics Online instance, you must first download the Microsoft Dynamics 365 Report Authoring Extension, which will extend the functionality of BIDS. With this extension, you can write FetchXML-based queries similar to how you would write SQL-based queries in traditional SSRS (SQL Server Reporting Services) reports.

The steps to install the Report Authoring Extension are as follows:

1. Make sure you have the prerequisites installed, which are Visual Studio and the Business Intelligence Development Studio—a downloadable add-on for Visual Studio (see Figure 4.13 shows SQL Server with BIDS installed).

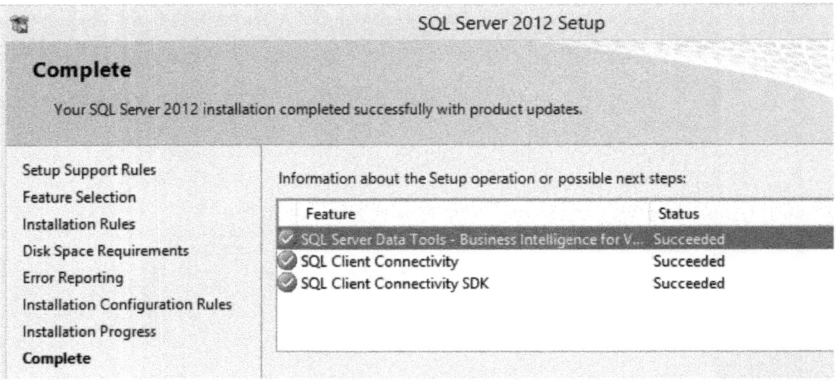

Figure 4.13: You can modify your SQL Server install with the Business Intelligence Development Studio if needed.

2. Download the CRM report authoring extension package from https://www.microsoft.com/en-us/download/details.aspx?id=50375 (or search for "Dynamics 365 Report Authoring Extension"). Figure 4.14 shows the download page.

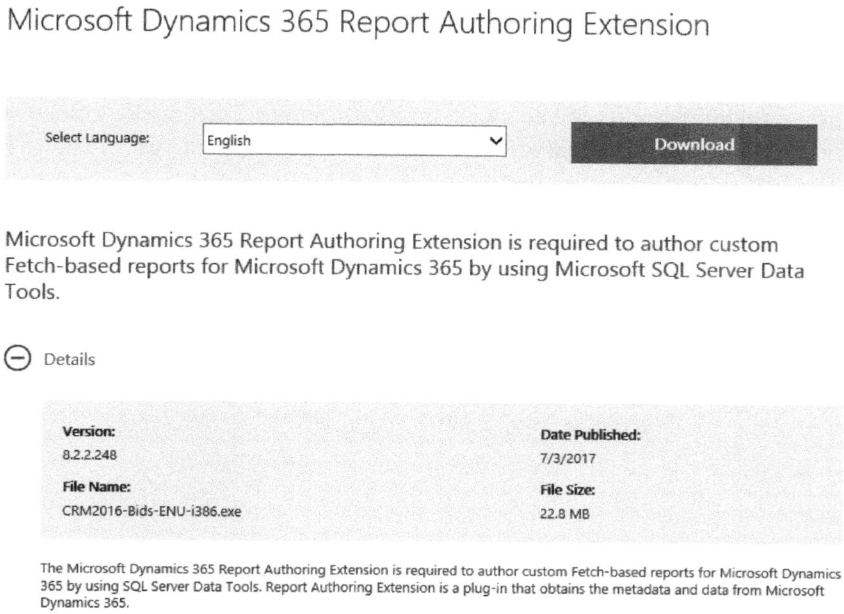

Figure 4.14: Downloading the BIDS extension for Visual Studio.

3. Run the EXE after it downloads. There will be a series of steps in the installation wizard that you'll need to click through, but assuming all of the prerequisites are in place, the default options should work. Once completed (Figure 4.15), you'll be able to create reports in SSRS.

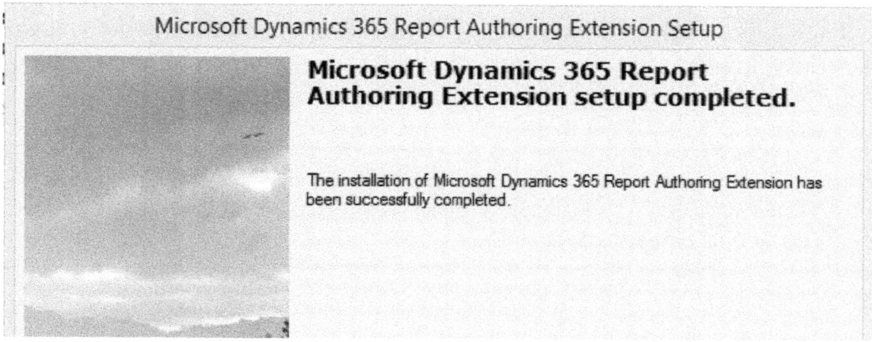

Figure 4.15: A quick install and confirmation of success.

Creating a Report Project

With your development environment prepared, you'll be able to create a report project and build reports. When you open Visual Studio, click the Report Server Project option under the Business Intelligence templates, as shown in Figure 4-16. Click OK to create the new project.

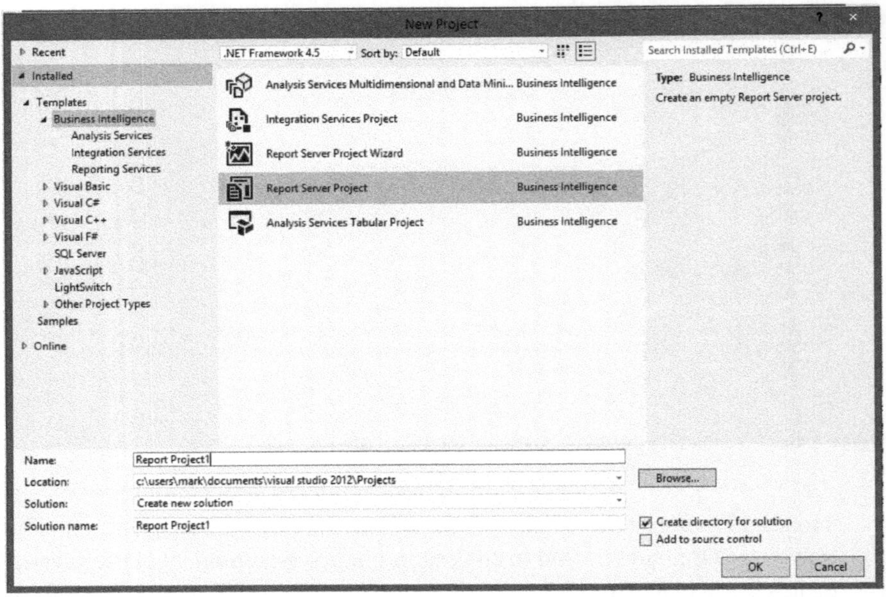

Figure 4.16: Creating a new report server project within BIDS.

In the Solution Explorer within BIDS, you will see a folder called Reports. Right click this and select Add New Report. This will open the Report Wizard shown in Figure 4.17. Once the splash screen appears, you can click Next.

You can use the Report Wizard to create your first report for any project. Once you have this initial report defined, you can simply copy and paste your original report as a template for future reports that will have the same "look and feel."

Figure 4.17: This is the report wizard within BIDS, as opposed to the report wizard you saw earlier.

The next screen in the wizard is where you can select your data source. We are going to look at reporting on information within a Dynamics 365 online environment, so the type of data source will be a Microsoft Dynamics 365 Fetch, as shown in Figure 4.18.

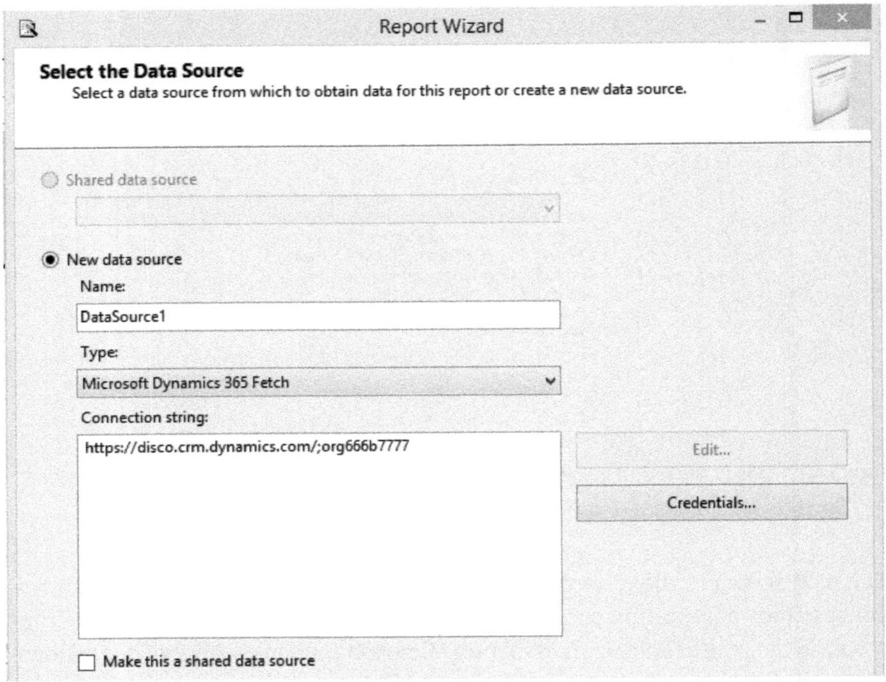

Figure 4.18: Connecting to your Dynamics 365 instance for your data source.

The connection string property requires a URL to be entered. In order to get the correct URL, first enter the value "https://disco.crm.dynamics.com/;". After the semi-colon, you must enter the unique organization name of your D365 instance. You can find this value by taking these steps:

1. Log in to your D365 instance and click on Settings in the main toolbar.
2. From the Settings drop down, click on Customizations.
3. On the Customization screen (shown in Figure 4.19), click on Developer Resources.

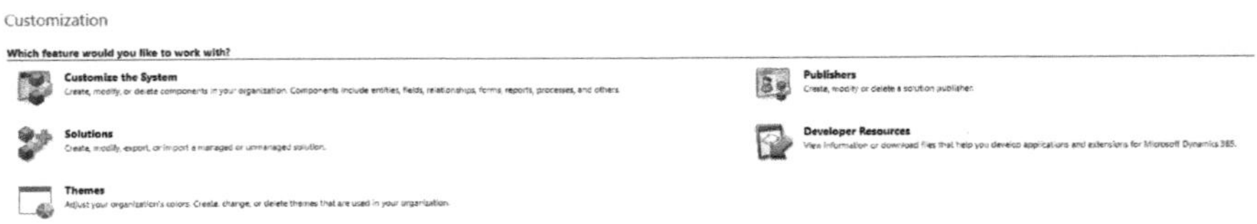

Figure 4.19: You can get to the Developer Resources from the Customization menu.

4. Within the Developer Resources screen, you will see your organization in the Unique Name field, which is under the Instance Reference Information header in the left-hand column (see Figure 4.20). Enter this value into the Connection string as shown in Figure 4.18.

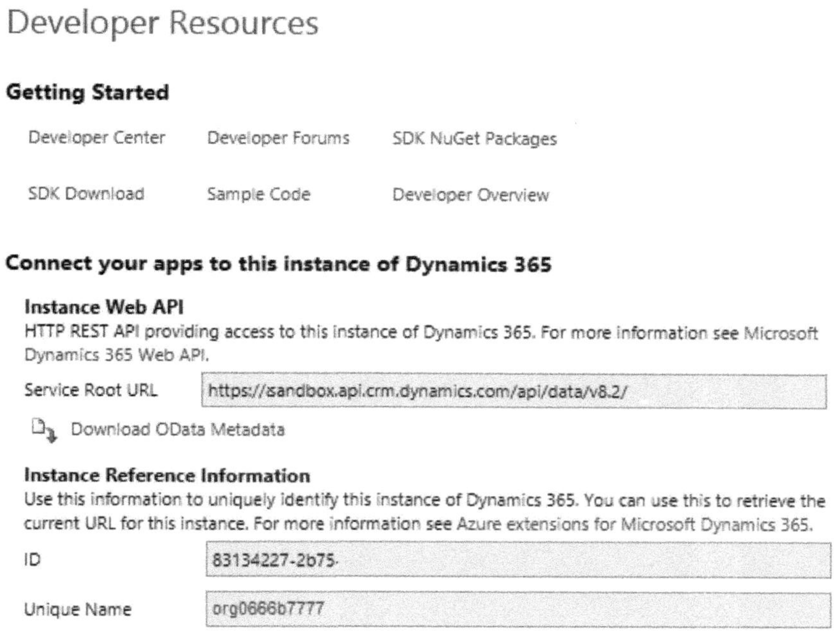

Figure 4.20: You will need the unique organization name information to configure your data source connection to CRM.

Once you have the connection string defined, click on the Credentials button (Figure 4.18). There are several options here, but for D365 connectivity the most likely option will be to "Use a specific user name and password," as shown in Figure 4.21. Enter the username and password that you would use when logging directly in to your D365 instance.

With the correct credentials entered, you can click OK within the Credentials window, and then Next on the main data source window in order to proceed to the next step of the wizard.

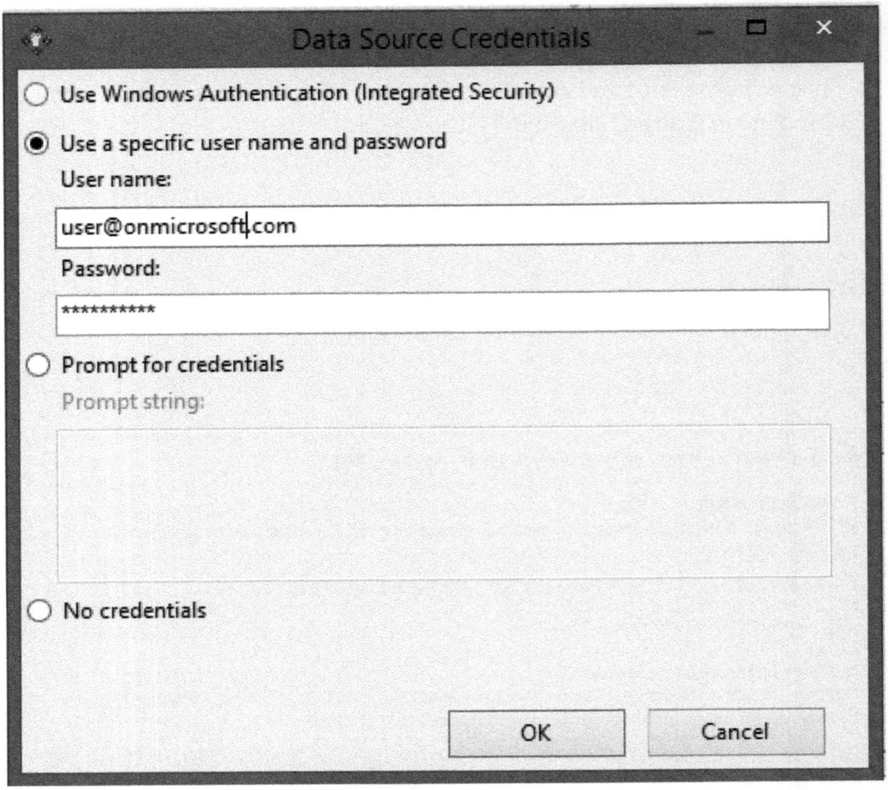

Figure 4.21: Enter your credentials.

The next screen is where the query that will retrieve the data your report is interested in can be defined. This query will be in FetchXML form, which is language specific to data retrieval from D365. There are many things that can be done with FetchXML and plenty of complexity associated with it, but for now you can enter a simple query by first going back to your CRM in the Advanced Find window. Create an advanced find query, such as pulling all leads with a certain type or all contacts with a specific last name—something basic that will return results. An example of an advanced find is shown in Figure 4.22.

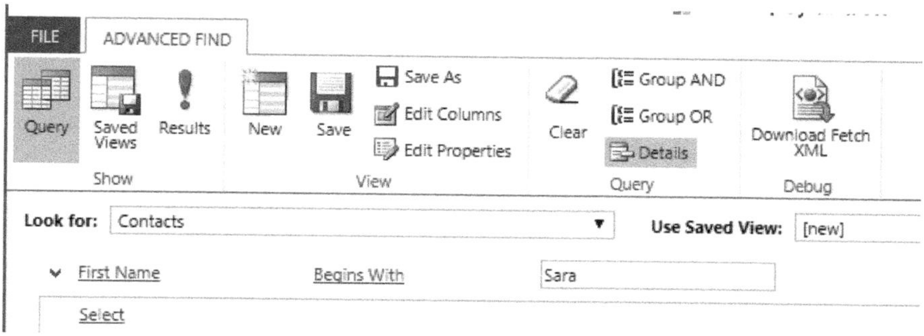

Figure 4.22: Advanced Find will generate FetchXML results for you.

Once you have defined your query within Advanced Find, click on the Download FetchXML button on the toolbar. This will download a file, which you can open in your browser, or in any text editor (like Notepad). The FetchXML for the query entered into the Advanced Find window will be displayed as XML, which can be copied and pasted directly into the BIDS query window. If you do open the FetchXML in an Internet Explorer browser window, make sure to right click the page and select View Source so that you get the clean text to paste. Note that the FetchXML in Figure 4.23 is identical in functionality to the Advanced Find query above in Figure 4.22. Click Next to advance to the next screen.

In the sample shown here, only three fields are returned—fullname, telephone1, and contacted. Additional fields can be returned by either editing the column list in Advanced Find and re-exporting the FetchXML, or by manually editing the list with an <attribute name="[yourfield]"/> notation.

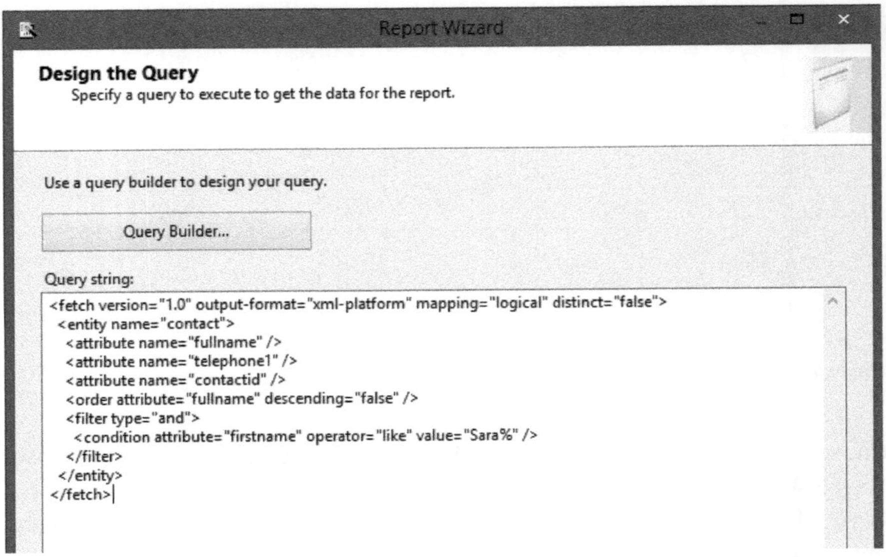

Figure 4.23: Pasting the exported FetchXML into the designer query window.

The next few screens are where the basic layout can be configured. If you want to go with the basic layout, then these will be acceptable—but typically, you will want to define your own custom layout. You can click Finish at any time if you are not interested in having the wizard give you a specific layout. In order to illustrate how to build out a custom report, click Finish in your wizard. A default report will automatically be created—Figure 4.24 is an example of what this will look like based on the FetchXML query shown earlier.

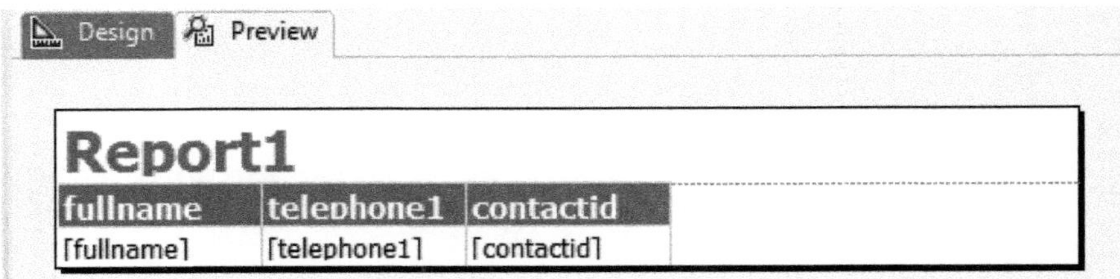

Figure 4.24: The default report layout that is generated.

Clicking on the Preview tab, the report will run and display a simple grid. Now let's look at expanding this grid's functionality. First, let's get a photo to display on each line. Let's assume that you have a photo uploaded for each contact. The photo is an attachment on a note within each contact's record. The photo will exist in the document body field of the note, while the subject line will have the value "Photo" in it. The note is an out-of-the-box CRM entity which has an actual entity name of "annotation." It is linked in this case to the contactId field (but this will be different depending on how you configure your attributes). The FetchXML in Listing 4.1 shows how this query would be constructed within the context of the initial query.

If you wish to edit the query, you can double-click the data set in the left-hand pane to open up the query again.

Listing 4.1. FetchXML Linking a Photo to a Contact

```
<fetch version="1.0" output-format="xml-platform"
mapping="logical"
distinct="false">
 <entity name="contact">
 <attribute name="fullname" />
 <attribute name="telephone1" />
 <attribute name="contactid" />
 <order attribute="fullname" descending="false" />
 <filter type="and">
 <condition attribute="firstname" operator="like"
value="Sara%" />
 </filter>
 <link-entity name="annotation" from="objectid"
to="contactid" visible="false"
 link-type="outer" alias="note">

 <attribute name="objectid" />
 <attribute name="documentbody" />
 <attribute name="subject" />

 <filter type="and">
 <condition attribute="subject" operator="eq" value="Photo"
/>
 </filter>

 </link-entity>
 </entity>
</fetch>
```

In your report, add a new column to display the photo by right-clicking the last column in the grid and selecting Insert Column (see Figure 4.25).

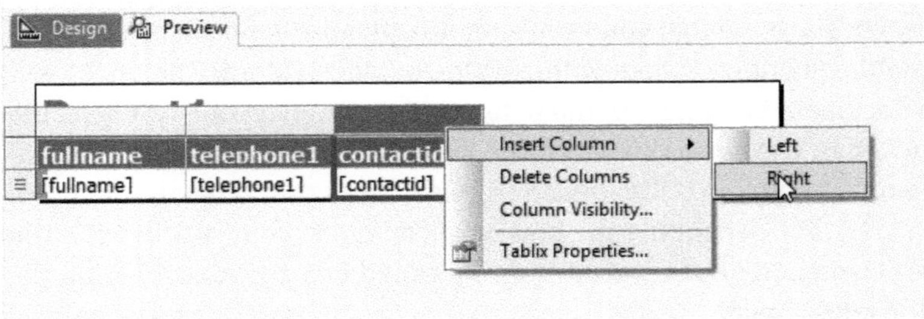

Figure 4.25: Adding a new column to the report.

Next, add a dynamic image to the column you just created. Right-click within the column detail (not the header) and select Insert and then Image from the menus, as shown in Figure 4.26.

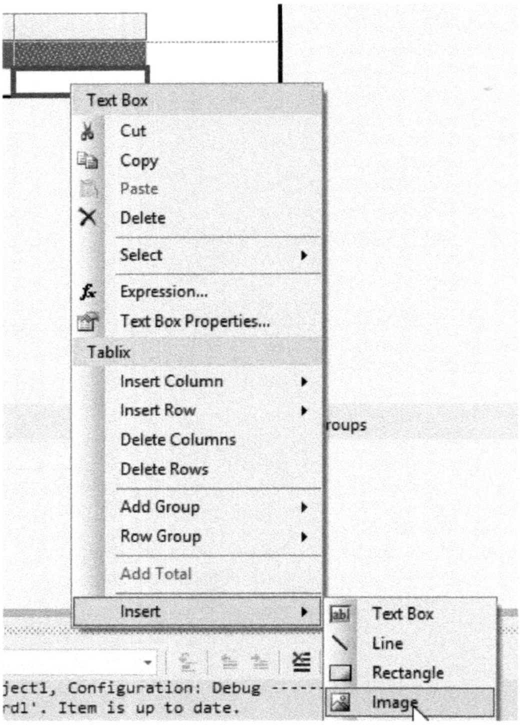

Figure 4.26: Configuring the column to display an image.

A window will open where you can configure the image to point to the field that you are returning in the FetchXML. You will want to select the image source and set the field and MIME type, as shown in Figure 4.27.

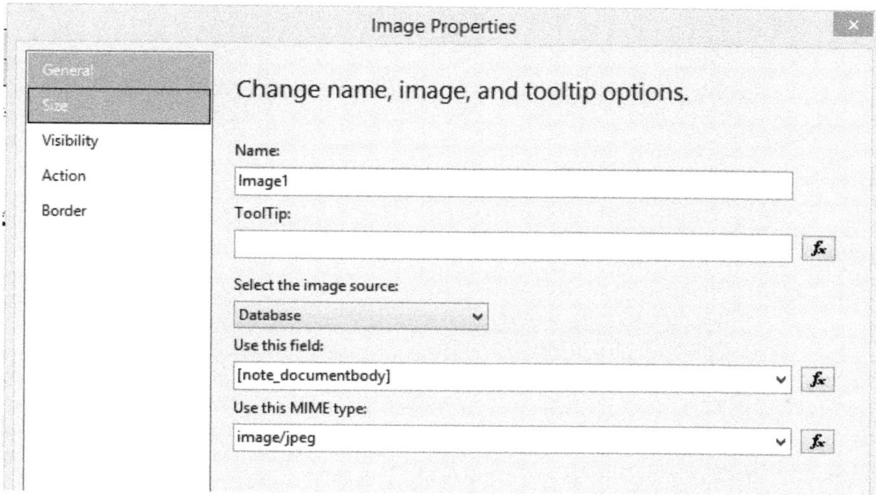

Figure 4.27: The image properties.

Once the image has been configured, you can preview your report by clicking on the Preview tab within the report designer. Once you have gotten a layout that you are pleased with, you can deploy your report to CRM using the following steps. Refer to Figure 4.28 for details.

1. Open a New report from with Dynamics.
2. Select Existing File for the Report Type Property.
3. Click Browse and browse to the RDL (Report Definition Language) file that you just built.
4. Give the report an appropriate Name and Description.
5. Set the Categorization properties. The most important step is to select the appropriate values in the Display In property. You can access your report from within a specific entity and from the main reports area, but you must define where it will be displayed using this property.
6. Click Save and Close. Your report will now be deployed.
7. If you want to edit your report, click the Edit button and re-upload the RDL file that you have been working with.

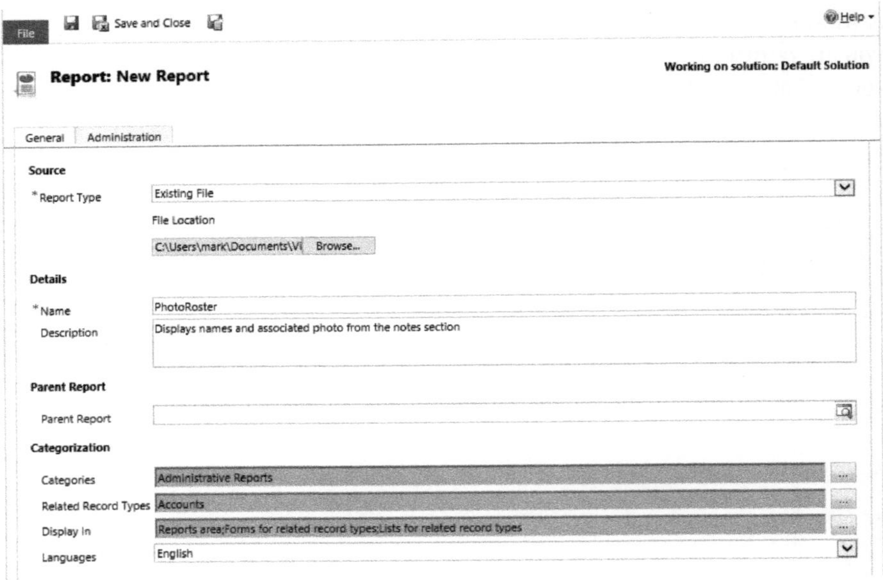

Figure 4.28: Uploading the RDL file you have created and setting the properties to deploy your report to CRM.

System Report Settings

One final item to note on reporting is that you can change the report categories in which you can organize reports. By default, these are:

- Sales Reports
- Service Reports
- Marketing Reports
- Administrative Reports

In order to change or add to this list of categories, open the Reporting tab within System Settings. You can navigate here by clicking Settings in the main ribbon, selecting Administration, then System Settings. Figure 4.29 shows the Reporting tab where these modifications can be made.

Figure 4.29: Modify the categories list in the Reporting tab.

Summary

The intention of this chapter was to give you an understanding of the options available to you for reporting within Dynamics 365. As you can see, the Report Wizard has very basic functionality and limited usage. BIDS, on the other hand, is a platform where you can develop complex logic and layout. What you have likely gathered is that while you can do whatever it is you need using FetchXML, it is far more cumbersome than traditional SQL. If you really want to use SQL, and you have an environment that requires it (such as advanced analytics and metrics), you do have the option of replicating data from your Dynamics Online instance down to a local SQL Server, however this would require data integration—which is covered in the next chapter.

Chapter 5
Importing Data

There are a number of approaches to loading data into Dynamics. This chapter will introduce several of these approaches. The first that will be covered is the use of the import wizard, a simple interface that allows a user to import data from a spreadsheet and map it to an entity. After that, we'll look at how to do the same work with Scribe Insight, which is a leading platform for Dynamics-related data integrations. We'll also take a brief look at other options for working with data.

Using the CRM Import Wizard

Users who have permission can import data into CRM from spreadsheets and flat files using an out-of-the-box import tool. This tool has basic functionality to map fields from the source file to fields in the targeted CRM entity. There is no ability to add formulas to transform data, or to do inserts and updates across multiple entities, but it does provide a simple interface to handle basic imports.

To import data, take the following steps. This demonstration will show you how to import new contacts into Dynamics 365.

1. Start by creating an Excel file. Define the columns and the data for each column. Figure 5.1 shows a list of four contacts in an Excel spreadsheet. The first row has the column headers in it.

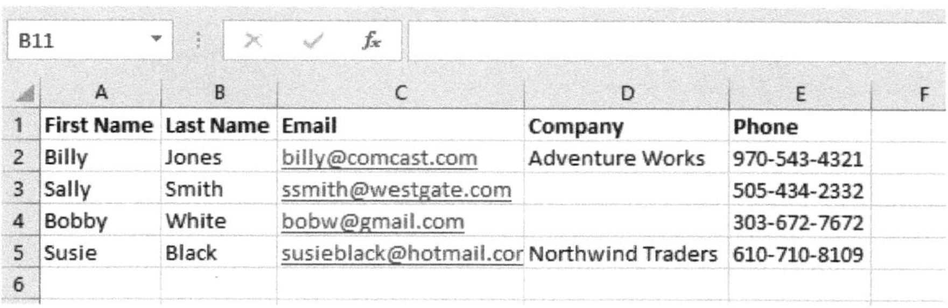

B11 | fx

	A	B	C	D	E	F
1	First Name	Last Name	Email	Company	Phone	
2	Billy	Jones	billy@comcast.com	Adventure Works	970-543-4321	
3	Sally	Smith	ssmith@westgate.com		505-434-2332	
4	Bobby	White	bobw@gmail.com		303-672-7672	
5	Susie	Black	susieblack@hotmail.cor	Northwind Traders	610-710-8109	
6						

Figure 5.1: A spreadsheet that will be used as the source data for the import.

2. Once this data is ready, save the file. Generally, working with a CSV file is easier, so when you save your file use the Save As button in Excel and save it

DOI 10.1515/9781547400065-005

as a CSV (see Figure 5.2). For reference, we'll refer to this file as "ImportDemo.csv".

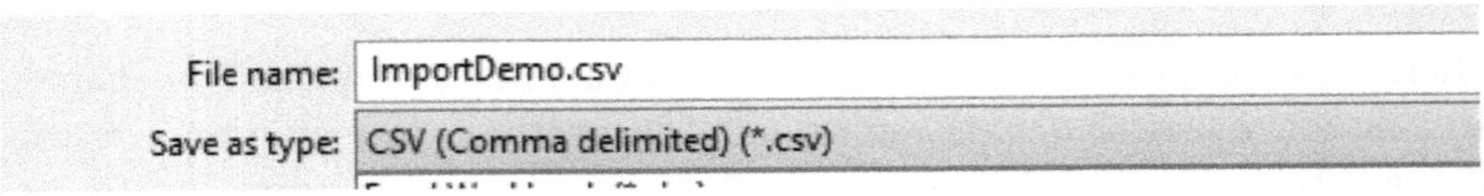

Figure 5.2: Saving the Excel doc as a CSV file.

3. In Dynamics 365, navigate to Contacts and click the Import Data button. The Upload Data File screen shown in Figure 5.3 will open. Here you can browse to the spreadsheet you created in the previous step. Click the Next button when you have your file added.

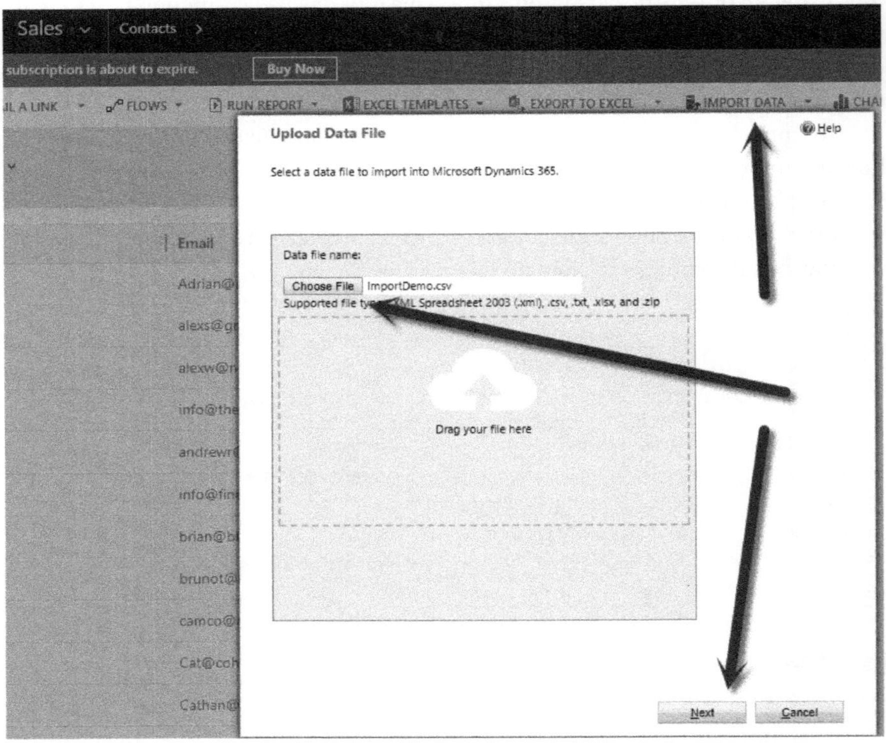

Figure 5.3: Choosing the file to upload in the import tool.

4. On the Review File Upload Summary screen, you can set the delimiters. Since you are using a CSV file, the default settings will work (shown in Figure 5.4). The Field delimiter property is set to Comma, but you could use several other options to separate your columns. The Data delimiter is set to Quotation mark. If, for example, you had a column called "Full Name", and you wanted it to read "Smith, Sally", then the import tool will look for the double quotes and not count the comma within that value as a field delimiter. Click Next on this screen to proceed.

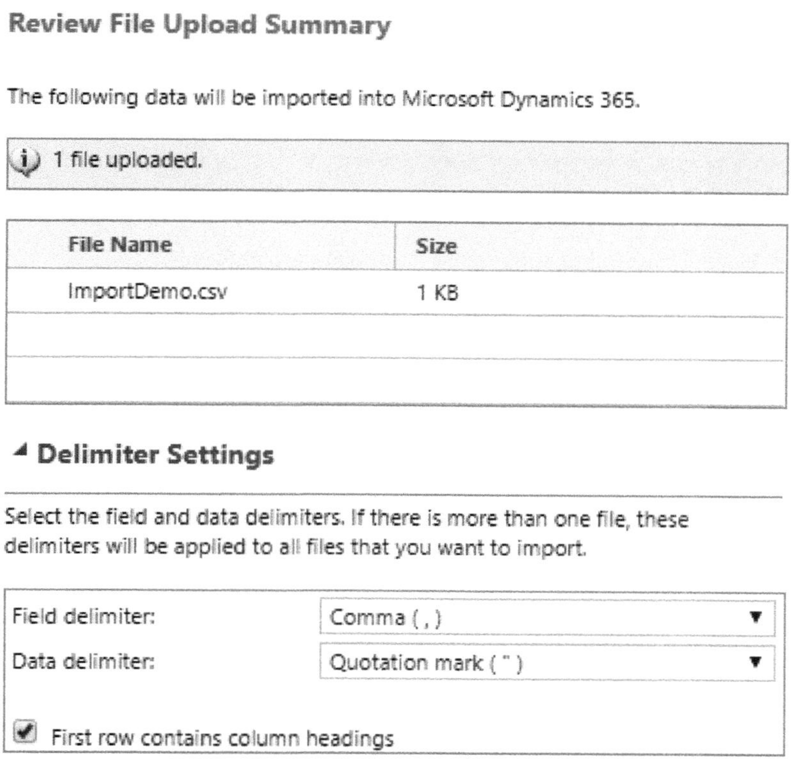

Figure 5.4: Setting the delimiters for the import.

5. Select Data Map is the next screen in the import wizard process. Here you will be able to point to a map template. Since this will be the first time you have imported this CSV format into CRM, you will use the Default (Automatic

Mapping) template, as shown in Figure 5.5. However, if you had imported this same format in the past, and had saved your mapping, the mapping template would be here. Click Next to proceed.

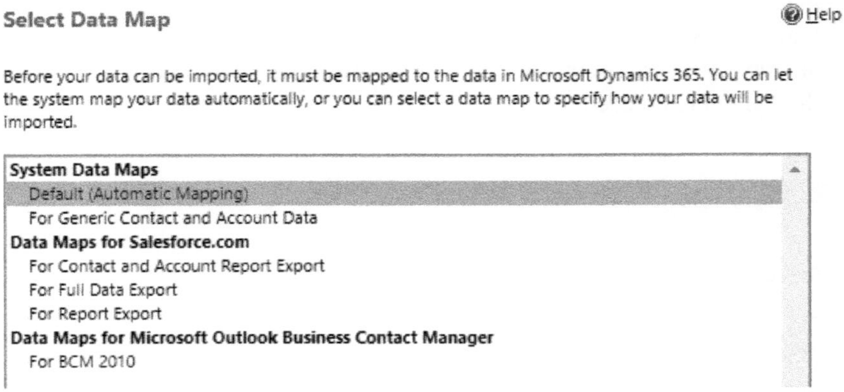

Figure 5.5: Specifying the map to use—using the default mapping for now.

6. The Map Record Types screen requires that you indicate what entity you are mapping this data into. For this example, we are mapping to the Contact entity. Select this entity (see Figure 5.6) and then click the Next button.

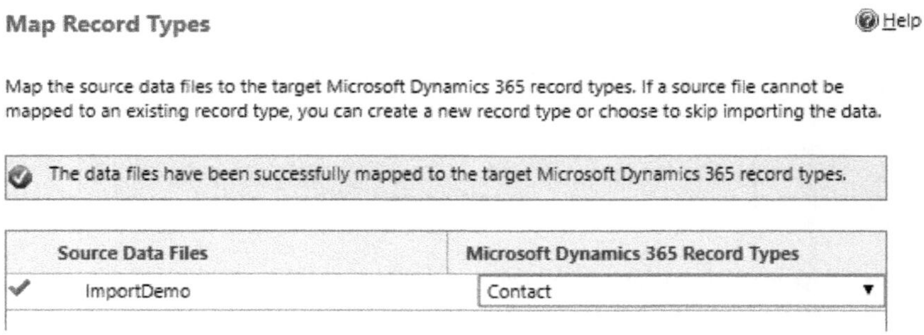

Figure 5.6: Selecting Contact as the targeted entity.

7. You are now on the Map Fields screen. This is where you will do your field mapping. Some of the fields will automatically map if the column name and the target field name in the CRM entity are the same. Others will require you

to select the field you are mapping to. Figure 5.7 shows all of the fields except for Company mapped. Company will be a lookup field, so we will look at that one separately in the next step.

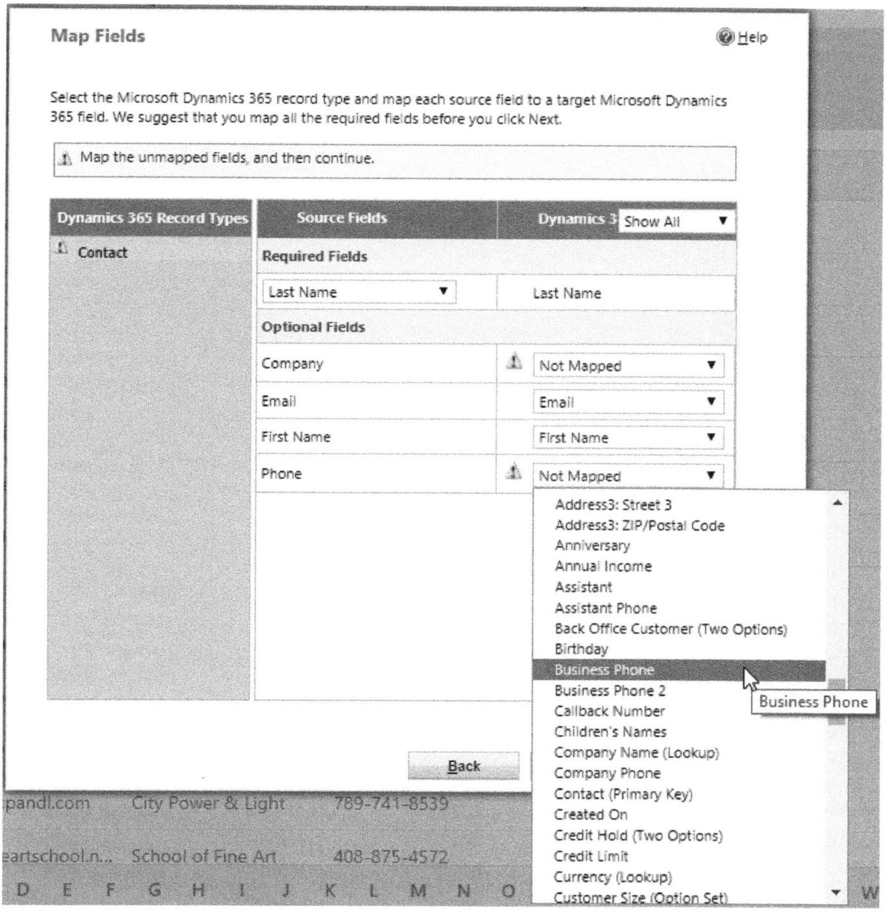

Figure 5.7: Mapping the simple text fields.

8. Now click on the Company drop-down menu. In this case, we will map it to the Company Name, which is a Lookup on the Account records. As you can see in Figure 5.8, selecting the Company Name (Lookup) option from the drop-down causes a Lookup reference section to appear on the screen. Here you can tell CRM what entity you are looking for and what field within that

entity should be used to do the comparison. In the cases of this example, the values in the Company Name column in the source CSV file will match exactly the values in the account record's Account Name field. Click the check boxes next to the Account Name and Full Name fields, then click OK.

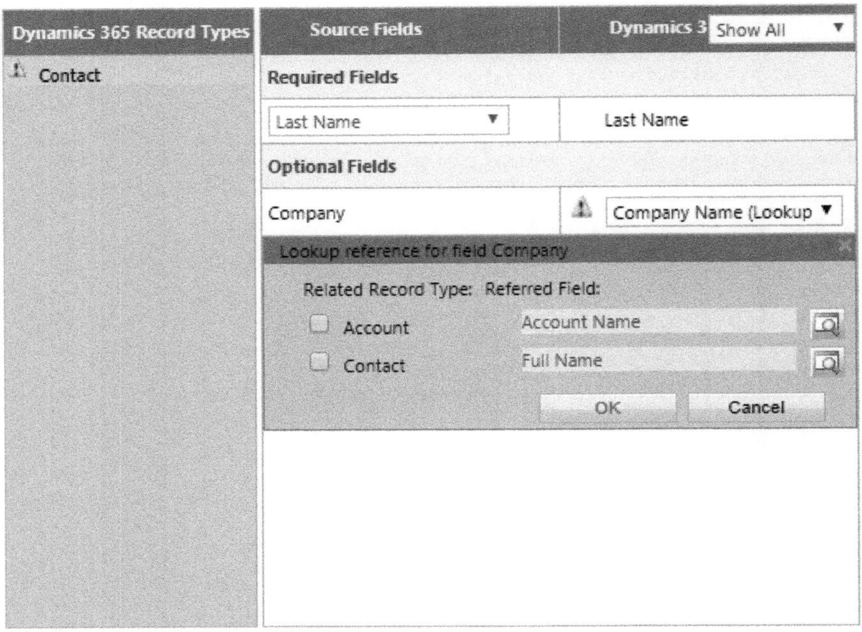

Figure 5.8: Mapping a lookup field.

9. When your mapping is complete, click on the Next button. A summary screen will display. Click Next again. This will show the final screen in the import wizard, which will allow you to determine several things.
   - Allow Duplicates—Generally, you will want this set to No to ensure you don't end up with a bunch of duplicate records. There may be cases where you do want to import the same data more than once, in which case you would set this value to Yes.
   - Select Owner for Import Records—By default, this will be set to the user that is currently logged in to the browser session, but can be set to someone else if needed. Often imports of data will be set to a system administration account.

- Data Map Name—If you would like to be able to reuse this map again, give it a name here.

10. When you are ready, click the Submit button. The final screen will show. From here, you can click the “Imports” link (see Figure 5.9) to see the status of your import, or you can click the Finish button.

Figure 5.9: You can click on the Imports link to see the status of the import you just ran.

Reviewing Data Import Status

You can always check the status of your imports as they are running or after they have completed. Clicking the Imports link as noted above will allow you to see your imports. You can also navigate here by clicking on Settings and then selecting Data Management from the main toolbar. Then click the Imports link (see Figure 5.10).

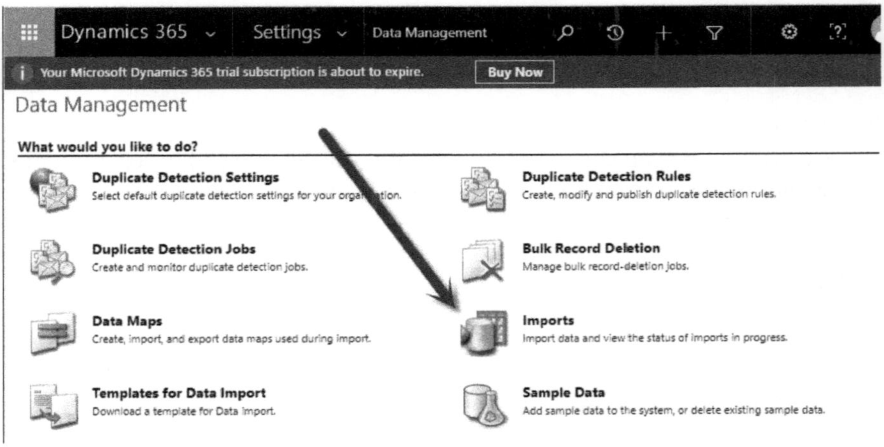

Figure 5.10: Access import status from the Data Management tab.

When you have the Imports window open, you will be presented with the default view of "My Imports." This will show the status of any imports you have run at any time. The information displayed includes the number of records processed and the number failed. In Figure 5.11, you will see the that the ImportDemo.csv file imported successfully.

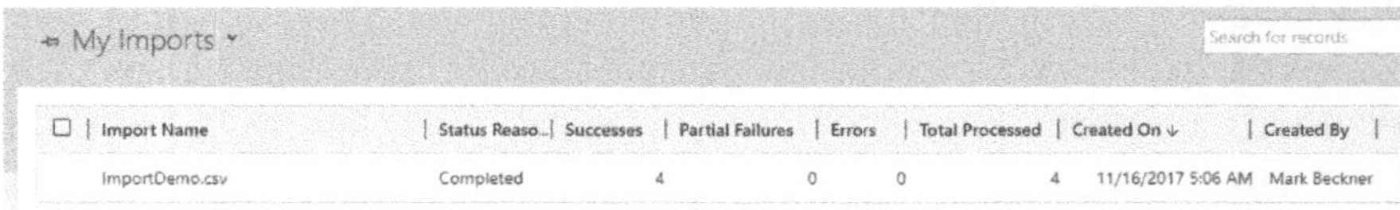

Figure 5.11: The list of imports and their status.

Clicking on the name of the import will open a new window where you can get details about your import. You can click on Contacts Fully Imported to see the list and open them in CRM (refer to Figure 5.12). If, for example, your import ran into errors, you could click on the Failures link to get a list of errors that occurred.

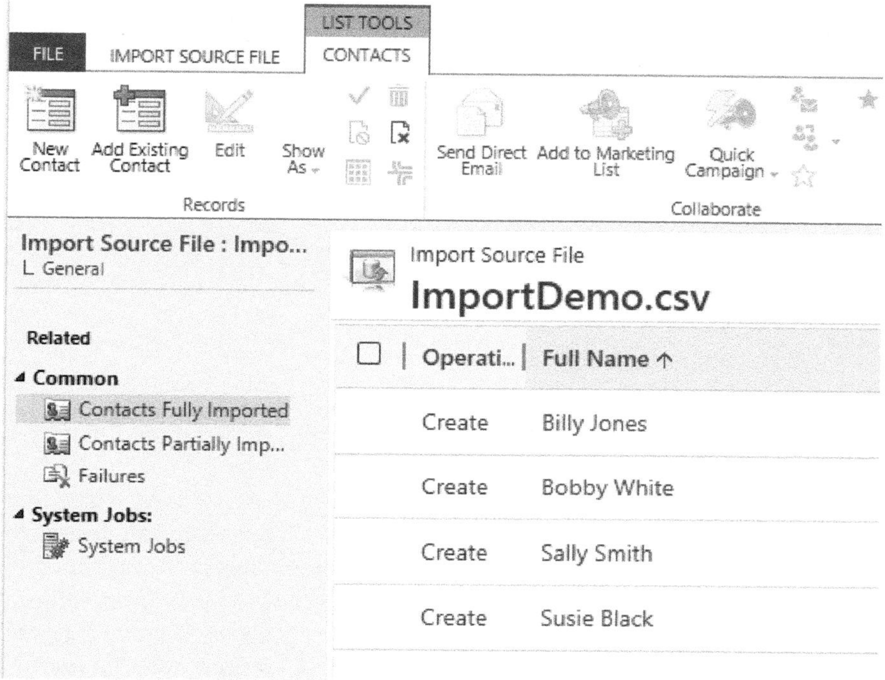

Figure 5.12: You can see the actual data loaded in the import status for an individual import.

Deleting Imported Data

A record of imported data will remain as long as you want it in Dynamics. This allows you to see who is importing data, how successful it is, and what records were loaded. If you have a bad import of data, and want to clear something out that has a large number of records, you can use the Bulk Record Deletion option on the Data Management page (Figure 5.13).

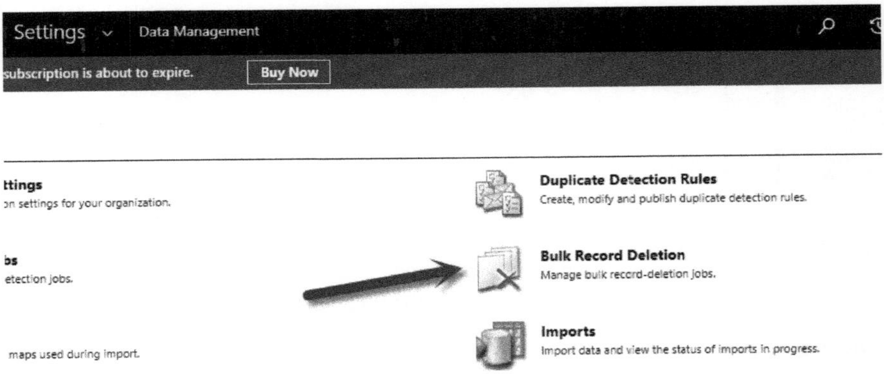

Figure 5.13: You can initiate a bulk deletion of data.

Clicking Bulk Record Deletion opens the Bulk Record Deletion page, which lists previously run job names. Click New to open a wizard where you can determine what records to delete. In Figure 5.14, the search criteria has been configured to delete the four records that were just imported. Once you have your query set, you can click the Preview Records button to ensure that your query is going to delete the records you are really after. If all looks good, you can proceed with the wizard.

If you only have a handful of records, you can use Advanced Find to delete. If you are dealing with hundreds of records or more, then you will want to use the Bulk Delete functionality.

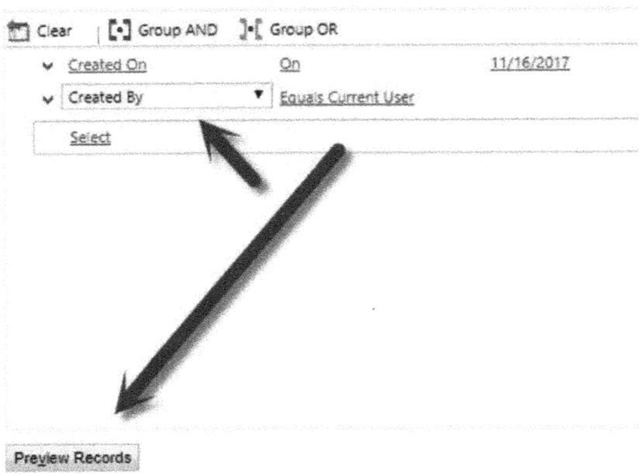

Figure 5.14: Specifying the criteria for deletion.

When working through the bulk deletion of data, you'll be able to set the schedule of the process. In this case, we'll set it to run at the current time (see Figure 5.15) and will send out an email when the process has completed.

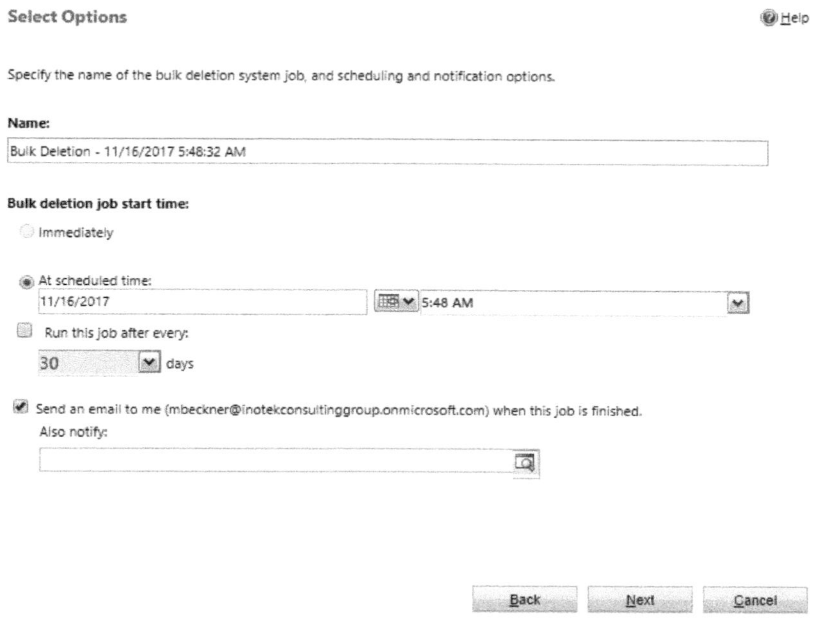

Figure 5.15: Scheduling the bulk deletion.

Clicking Next will allow you to complete the wizard and the job will be scheduled. Back on the main Bulk Deletion screen, you will see a list of jobs that have been scheduled, along with their status. The bulk delete will run in the background, and often takes a few minutes to initiate. As you can see in Figure 5.16, the job is in a Waiting status, which means it is scheduled to run but has not yet initiated.

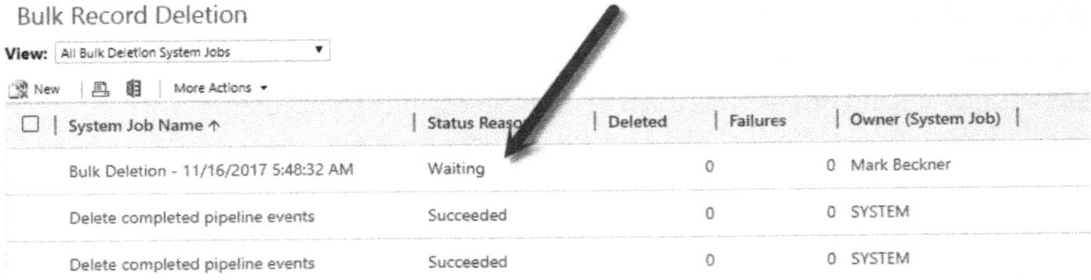

Figure 5.16: Checking the status of the bulk deletion.

Using Data Import Maps

When you have saved a data map, you can use it again for future imports. All data maps can be accessed from the Settings menu in the Data Management area (see Figure 5.17).

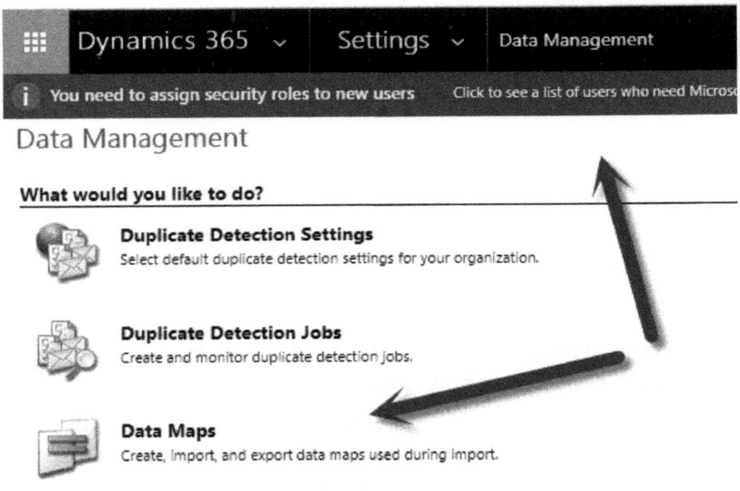

Figure 5.17: Access Data Maps.

Click the Data Maps link to open a view into the maps that are available to you. As you can see in Figure 5.18, there is one map that was created earlier in this chapter. Clicking on it will open a window with details about when the map was created—but, surprisingly, you can't actually edit the map. In cases where a map is complex, and you would like to know what the mapping was, you will have to export the map. This can be done by clicking the Export button on the toolbar.

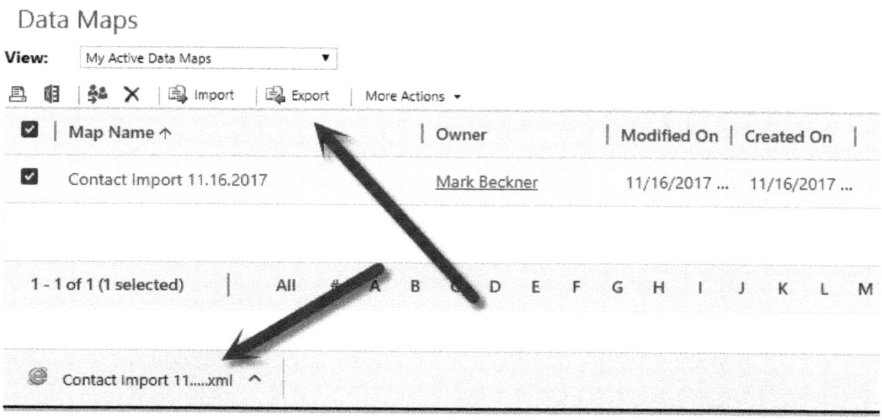

Figure 5.18: Exporting a data map.

When you click the Export button, you will get a downloaded XML file representing the map. If you are in Internet Explorer, check the lower left corner of your screen for the downloaded file. Clicking on this will open the XML in Internet Explorer. The XML code shown in Listing 5.1 is the actual map exported. It shows all of the attribute mappings, and you could, if needed, piece together what this map does and recreate it on a future import. You can also import this map, or make changes to the XML and import it as a new map.

Listing 5.1. Exported Map in XML

```
<Map Name="Contact Import 11.16.2017" Source="Import">
<Description>Data map created automatically using
map</Description>
<EntitiesPerFile>Single</EntitiesPerFile>
<EntityMaps>
<EntityMap TargetEntityName="contact"
SourceEntityName="ImportDemo" Dedupe="Ignore"
ProcessCode="Process">
<AttributeMaps>
<AttributeMap>
<SourceAttributeName>Email</SourceAttributeName>
<TargetAttributeName>emailaddress1</TargetAttributeName>
<ProcessCode>Process</ProcessCode></AttributeMap>
```

```
<AttributeMap>
<SourceAttributeName>Last Name</SourceAttributeName>
<TargetAttributeName>lastname</TargetAttributeName>
<ProcessCode>Process</ProcessCode>
</AttributeMap>
<AttributeMap>
<SourceAttributeName>Company</SourceAttributeName>
<TargetAttributeName>parentcustomerid</TargetAttributeName>
<ProcessCode>Process</ProcessCode>
<LookupMaps>
<LookupMap>
<LookupType>System</LookupType>
<LookupEntityName>account</LookupEntityName>
<LookupAttributeName>name</LookupAttributeName>
<ProcessCode>Process</ProcessCode>
</LookupMap>
</LookupMaps>
</AttributeMap>
<AttributeMap>
<SourceAttributeName>Phone</SourceAttributeName>
<TargetAttributeName>telephone1</TargetAttributeName>
<ProcessCode>Process</ProcessCode>
</AttributeMap>
<AttributeMap>
<SourceAttributeName>First Name</SourceAttributeName>
<TargetAttributeName>firstname</TargetAttributeName>
<ProcessCode>Process</ProcessCode>
</AttributeMap>
</AttributeMaps>
</EntityMap>
</EntityMaps>
</Map>
```

Using Scribe to Import Data

The basic functionality that the out-of-the-box import tool provides is often not robust enough to deal with real data imports that are happening in larger organizations. You may have integration between systems that import data into CRM on recurring schedules, or you can do things on a one-time basis.

Most organizations that have any type of integration or large importing requirements around Dynamics use Scribe. This section is intended to give you an idea of how Scribe can be used and what it looks like in action, so that you can understand the level of effort required to use it in comparison with the out-of-the-box import tool.

To illustrate what a Scribe import of data would look like, we'll use the same source data that we looked at earlier. In this case, we'll take the same spreadsheet of data and paste it into a SQL Server database. Figure 5.19 shows the data in a SQL table, and Listing 5.2 shows the SQL code to create the table.

FirstName	LastName	Email	Company	Phone
Billy	Jones	billy@comcast....	Adventure Works	970-543-4321
Sally	Smith	ssmith@westga...	*NULL*	505-434-2332
Bobby	White	bobw@gmail.c...	*NULL*	303-672-7672
Susie	Black	susieblack@hot...	Northwind Tra...	610-710-8109

Figure 5.19: The data that was in the CSV file earlier is now in a SQL table.

Listing 5.2. Scripted Table of Source Data for Scribe

```
CREATE TABLE [dbo].[CustomerList](
 [FirstName] [varchar](50) NOT NULL,
 [LastName] [varchar](50) NOT NULL,
 [Email] [varchar](50) NOT NULL,
 [Company] [varchar](50) NULL,
 [Phone] [varchar](50) NULL
) ON [PRIMARY]
```

Open the Scribe Workbench and create a new connection to your database where you create this table. Then point to the table and select the data from it, as shown in Figure 5.20. This will allow Scribe to use this as the source data for the import.

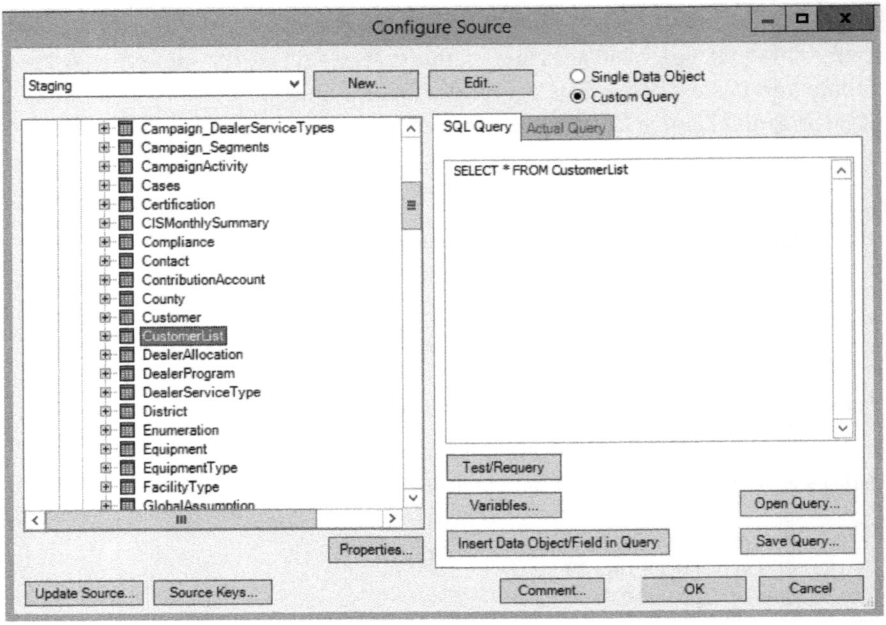

Figure 5.20: Setting the source data in Scribe.

Next, configure your target. This will be a connection to your Dynamics 365 environment, an example of which is shown in Figure 5.21. When you define your target, you select the entity that you want to write data to, and then you create an operation. In this case, we will select the Contact entity as the target in Dynamics, and will use an Update/Insert operation. This operation will first look to see if the record exists (based on the email address in this example), and if it does exist, it will update it. If it does not exist, it will insert it.

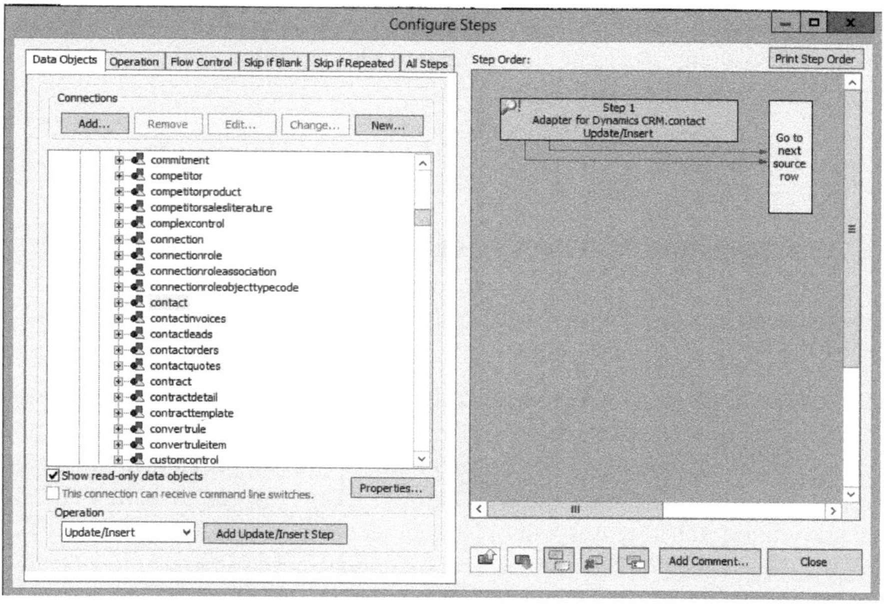

Figure 5.21: Setting the target and specifying the operation.

When the source and target connections have been defined, you will be able to map your fields from the source to the target in the main Scribe window. A field that has its value mapped to the target is a Data Link, while a field that is used to perform a lookup is a Lookup Link. In this example, the email address will both be a Data Link and a Lookup Link. If the record exists in CRM, the email address will be used to look it up. If it does not exist, then when the record is created, it will create it with the email address data. Figure 5.22 shows these links.

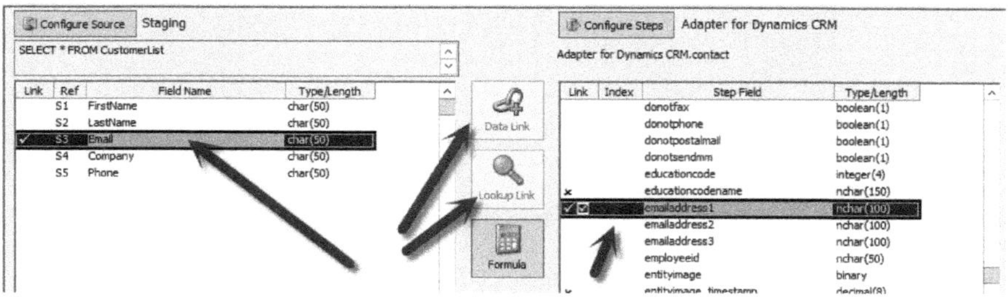

Figure 5.22: Mapping data.

Map the FirstName, LastName, and Phone across as Data Links. Once these are mapped, you should see something similar to what is shown in Figure 5.23 in the Data Formulas tab of Scribe.

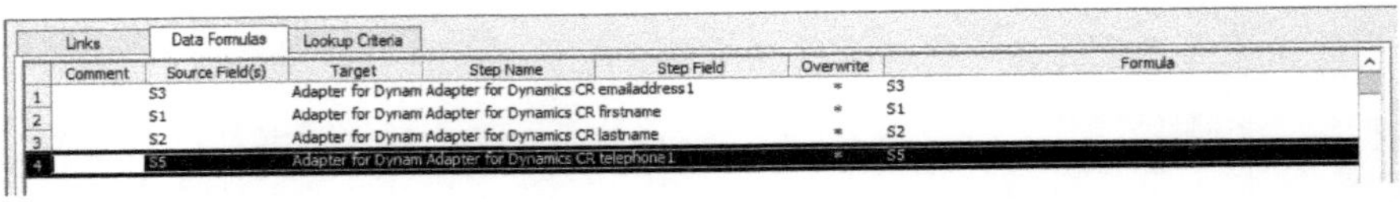

Figure 5.23: The Data Formulas tab shows that things have been mapped.

The only field left to map is the Company. This will require that we do a lookup in CRM to get back the GUID (Globally Unique Identifier) from the account. To do this in Scribe, create a new User Defined field by clicking on View in the menu and click User Variables. A window will appear where you can click the Add Calculated button. Clicking this will open another window where you can enter a DBLOOKUP command. This command allows you to specify the source field and the targeted entity to do the lookup against. Figure 5.24 shows the various settings for this lookup.

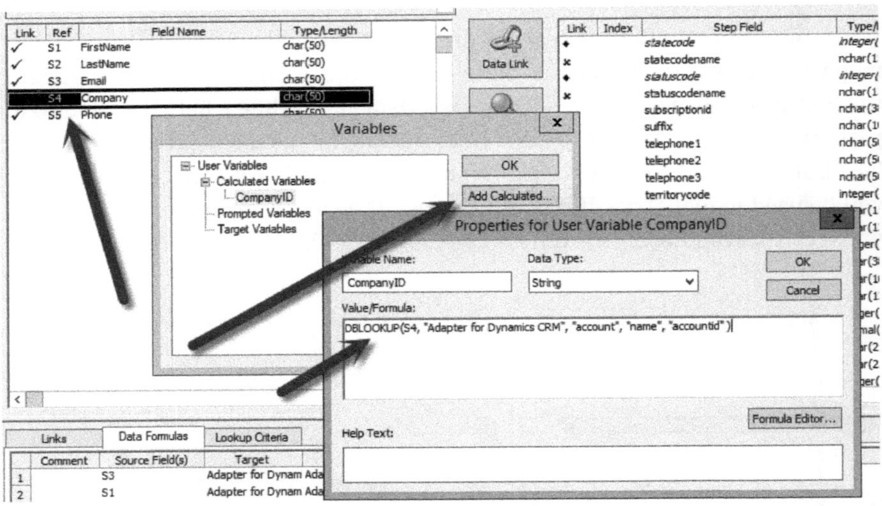

Figure 5.24: The DBLOOKUP allows for looking up a GUID.

Once the new user defined field has been created (we'll call it CompanyID), you can add it to the targeted field. You'll specify the account field in the target. Click on this field and press the Formula button and type in CompanyID. Figure 5.25 shows how this is set.

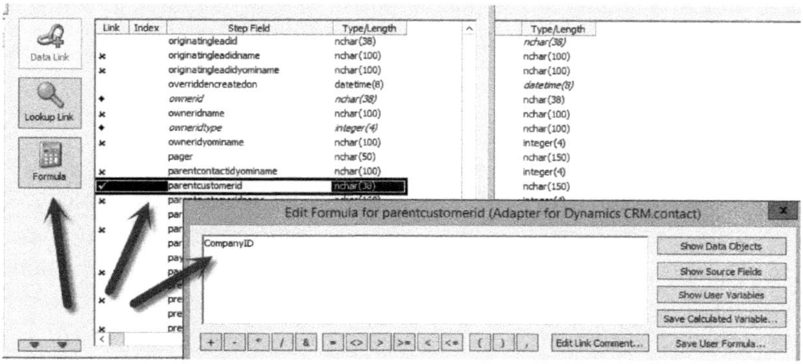

Figure 5.25: Adding the new user defined field to the target field.

This will complete the mapping for Scribe. You can now test the process by clicking the checkbox icon on the Scribe header. This will pull the source information for a record and show where an update or an insert will occur. Figure 5.26 shows the test window, with source data and outcome of the execution.

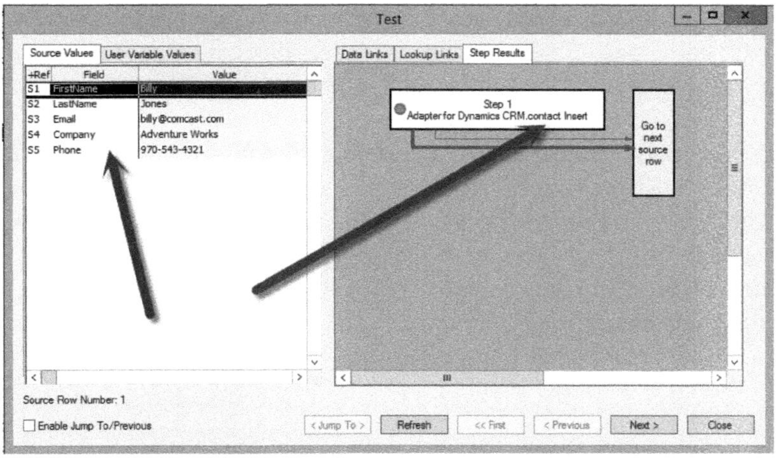

Figure 5.26: Testing the Scribe integration shows the expected result.

This brief introduction should give you an idea of what is involved in working with Scribe. While the steps are more involved for this specific example than using the out-of-the-box import tool, you can see that Scribe has far more functionality to allow for formulas, lookups, and other custom code that could never be done with the regular import tool. Scribe is very versatile and is fairly simple to ramp up and utilize. If you have complex imports, ongoing imports, or integration between systems, you will want to explore Scribe in more detail.

Data Integration

You will have many cases where you need to move data between Dynamics 365 and other systems. For example, you may want to create an account automatically in D365 when the account is created in your external accounting system. There are a number of options available for data integration, which include the following:

- SQL Server Integration Services (SSIS)—There is an adapter for SSIS that allows it to communicate directly with Dynamics.
- Scribe—Scribe has two versions, On Premise and Online. With either, you have the ability move data to and from Dynamics either on a schedule or as records are written. Scribe Online also has a replication feature, which will allow you to point to your Dynamics Online instance and pull down whatever entity's data you want into a local SQL Server database.
- BizTalk—Some larger organizations use BizTalk to move data between all of their environments and use it to push and pull data through the CRM API.
- Custom .NET code—You can always write C# code to interact with Dynamics.
- Other third party tools—There are several vendors that have mapping and integration tools specific to Dynamics 365.

Getting into the details of architecting and developing an integration platform or solution to interact with Dynamics is beyond the scope of this book but, as a system administrator, you do need to become familiar with the tools that are available to you for importing and exporting your data. The CRM-specific options of the out-of-the-box importer and exporter are not robust enough to allow you to build out automated integrations.

Summary

We've covered the basics of importing data into CRM, and you should have clarity on the options that are available to you. You should also understand the basic level of complexity associated with these imports. You will have the option to allow users to import their own data (using the out-of-the-box importer) or disabling this functionality. It is often useful to let a small set of power users have access to import data, but opening it up to too many users can often lead to dirty data and failed imports.

Chapter 6
Business Process Flows and Business Rules

Defining and enforcing consistent business processes and rules is one of the main reasons people use Microsoft Dynamics 365. Consistent processes help make sure people using the system can focus on their work and not on remembering to perform a set of manual steps or multiple rule scenarios. Processes and rules can be simple or complex and can change over time. This chapter outlines the basics for implementing both processes and rules so that you understand the level of effort required to configure and use these in your own environment.

Business Process Flows

Key points to know about *business process flows* (sometimes referred to as BPFs) are:

- Use a business process flow when you want staff to move through the same stages and follow the same steps in interacting with a customer. For example, use a business process flow if you want everyone to handle customer service requests the same way, or to require staff to gain approval for an invoice before submitting an order.
- Microsoft Dynamics 365 includes several ready-to-use business process flows for common sales, service, and marketing tasks that you can use with little or no changes required. Or you can create your own.

You can take the following steps to set up a business process flow in Dynamics:

1. Navigate to Customizations via the main Settings menu on the toolbar and click Customize the System.
2. On the Default Solution screen that opens, click on the Processes tab. A list of all existing processes will display, similar to what is shown in Figure 6.1.

DOI 10.1515/9781547400065-006

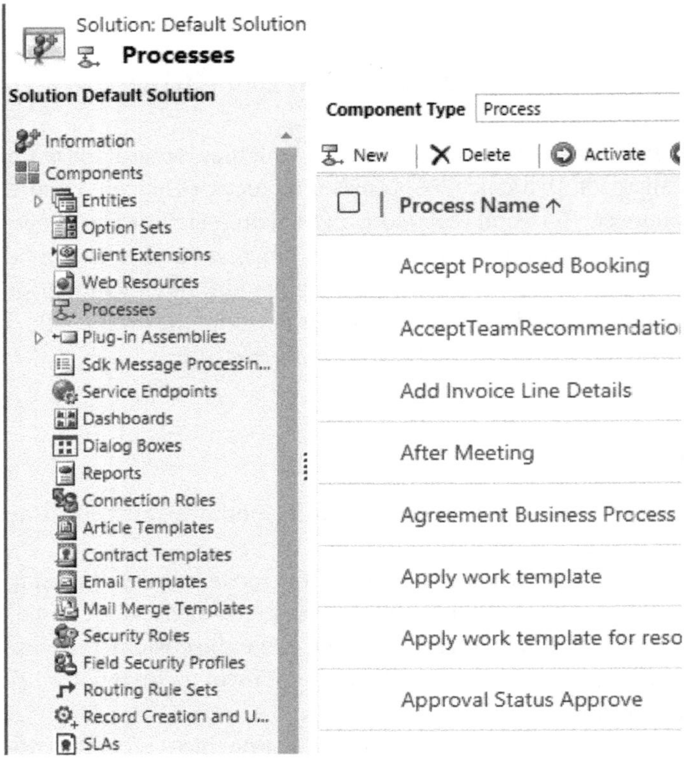

Figure 6.1: Accessing the business process flows from the Default Solution screen.

3. Click the New button and the New Process screen will open. Give the process a name and then select Business Process Flow from the Category property.
4. Next, select the entity that the business process flow will be associated with. In the case of this example, we will use the Account entity as shown in Figure 6.2. Once selected, click the OK button to proceed.

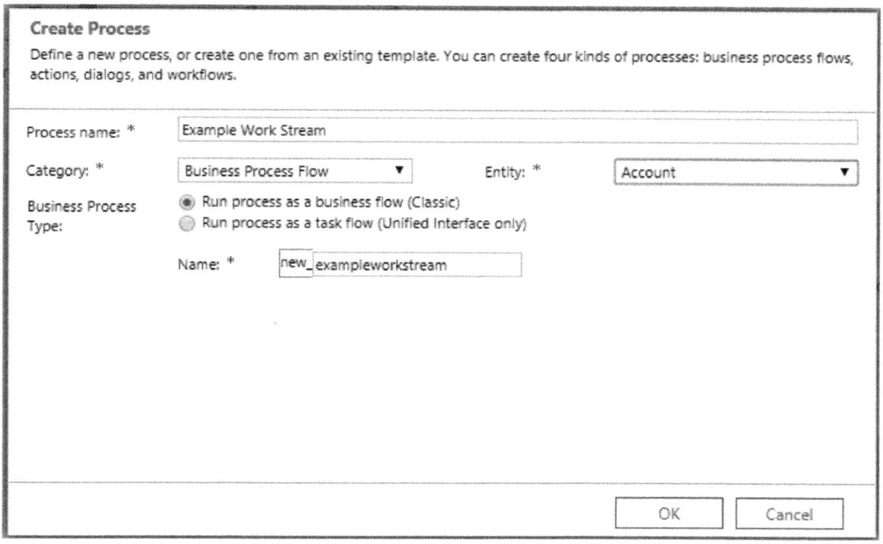

Figure 6.2: Creating a new business process.

5. After clicking OK, a new window will open which is the Business Process Flow development screen (see Figure 6.3). It provides for a variety of configurations.

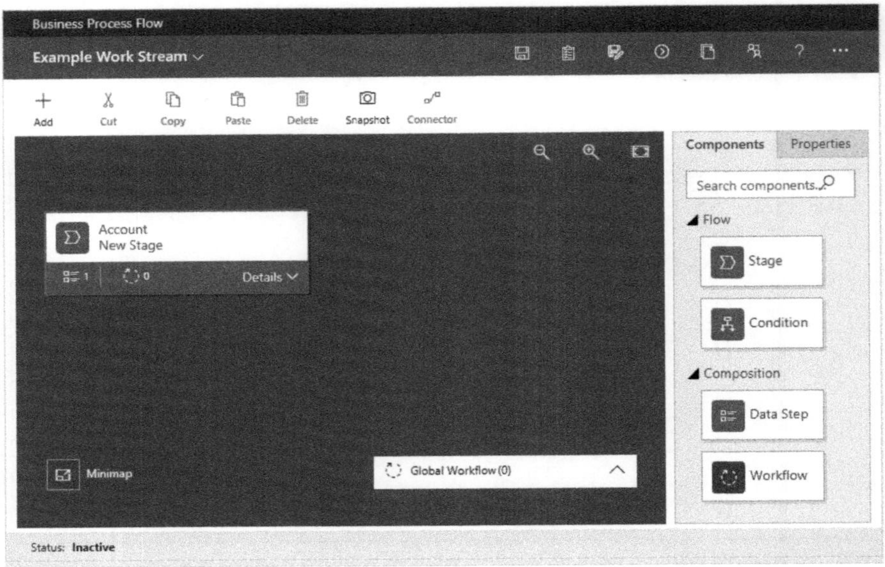

Figure 6.3: The business process flow development window.

6. The first step is to define what fields are going to be available at the first step of the process, and what that first process step will be named. You should have a default stage tile for the Account (called New Stage) that is available when you start—if not, drop a Stage tile onto the work surface.
7. Set the name of this stage by going to the Properties tab and modifying the Display Name property. This will be the name of the chevron that is added to your account's BPF ribbon once this process is complete.
8. Set the Category property of the tile. You can choose the category that is most appropriate, but for this example we will select Develop, as shown in Figure 6.4. Make sure to click the Apply button when you have set this tile property.

There is also a link to add business rules to a stage. Business rules will be covered later in this chapter. Once defined, they can be added into your business process flow.

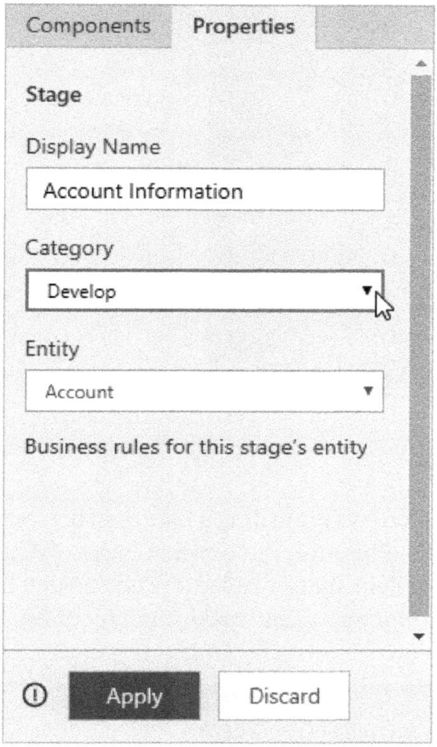

Figure 6.4: Setting the properties on a BPF stage.

9. Click on the Details drop down on the tile. This will allow you to configure what fields will show up in the ribbon of this stage. These fields are promoted to the top of the screen so that users can easily access and set them without having to hunt for them within the standard form. As you can see in Figure 6.5, a step where the Account Name field has been set to Required has been added to the Account stage.

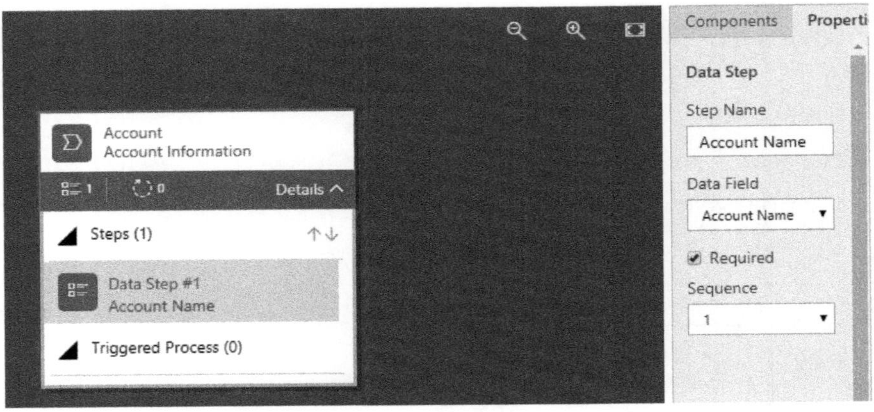

Figure 6.5: Configuring a step in a BPF.

10. You can add additional fields as well—one way is to drag a Data Step tile from the Components tab and drop it on the Plus area, as shown in Figure 6.7. In this example, we will add a second field that is called Account Number (based on accountnumber in the account entity) and make it not required.

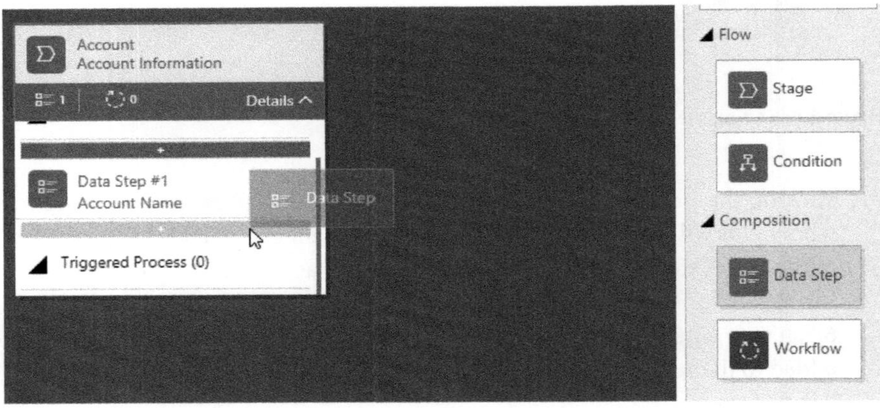

Figure 6.7: Dragging and dropping a Data Step onto the design surface.

11. Next, add a new Stage by dropping a Stage tile on the development surface. This Stage will reference a different entity (the Opportunity) to illustrate how data can be bubbled up onto the form from various related entities. In Figure

6.8, you can see that the Entity field is set to Opportunity and the relationship is defined as being on the Account.

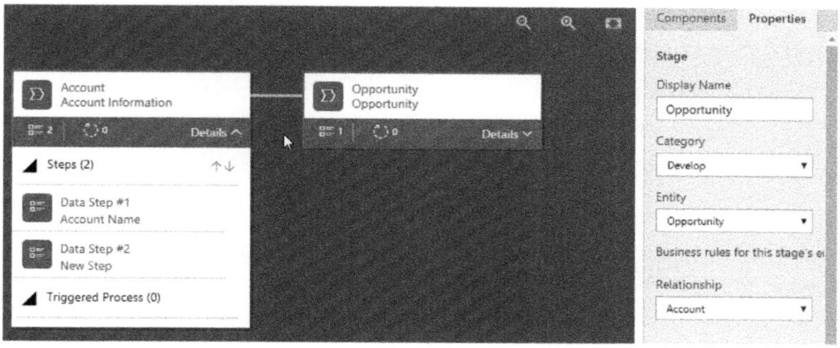

Figure 6.8: Adding a new stage.

12. At this point, we can save this BPF by clicking the save button on the ribbon at the top of the BPF development window.
13. You can validate what you have configured by clicking the Validate button (next to the Save button on the toolbar). This will show if there are any errors, and print the error information on the development surface and next to the properties that have issues. As you can see in Figure 6.9, there is one error on a field that was added to the surface, but was never saved. To resolve the issue, make the change to the field and click the Apply button. Try to Validate the process again. If everything looks accurate, you will see "Validation Successful" printed at the top of the screen.

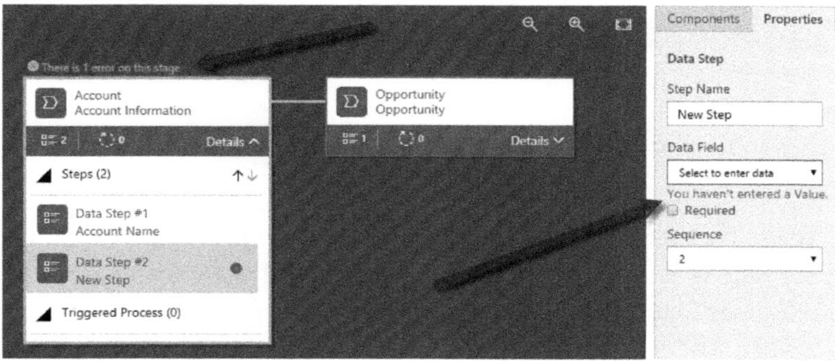

Figure 6.9: Validating a business process flow.

14. Now, click the Order of Business Process Flows in the toolbar (you will need to hover over the icons to see the value, but it is the one that has two file icons on top of each other). This will open the window shown in Figure 6.10.

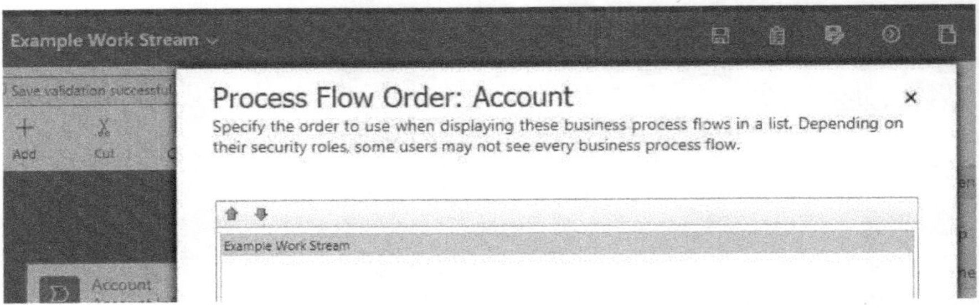

Figure 6.10: Setting the order of execution when there are multiple flows.

If you ever have more than one BPF defined for an entity, CRM will allow you to access these flows from within the entity. For example, if we had two BPFs defined on the account entity, the first one in the list would show up by default in the account record. You can switch between BPFs once you are in a record by clicking on the Switch Process button, as shown in the ribbon for an account in Figure 6.11.

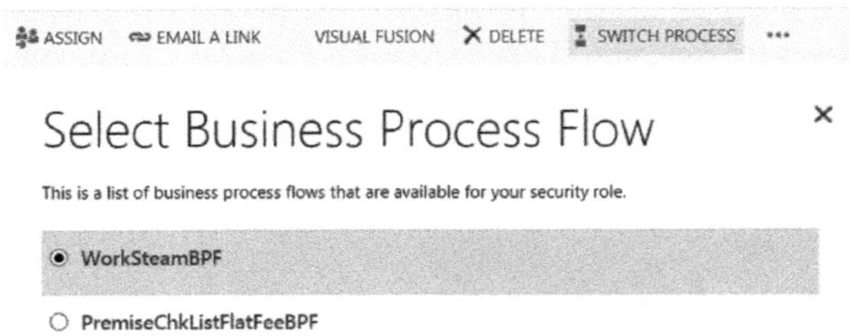

Figure 6.11: You can switch which flow is executing if there are multiple flows available.

15. You can assign the business process flow to a specific set of security roles by clicking the Edit Security Roles button. A window will open with the list of security roles (see Figure 6.12). You can click on a specific role and open the role for editing. There is a Business Process Flow tab where all of the BPF access permissions for that role can be set.

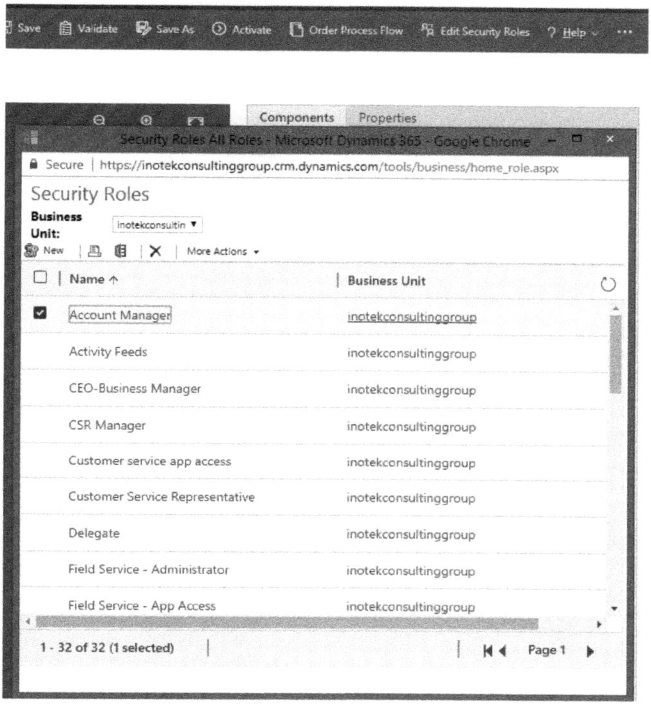

Figure 6.12: Roles can be given permissions directly on BPFs.

16. Once everything is configured and validated, you can activate the business process flow, which will add it to the targeted entity. Back in the process flow list, select the BPF you just created and click the Activate button. You can access all of the business process flows from the main Settings menu under Processes. Figure 6.13 shows where the Activate button is located. Notice that there is a column called Status which shows whether the process is Active or not (the one in this image is currently still in Draft and has not yet been activated).

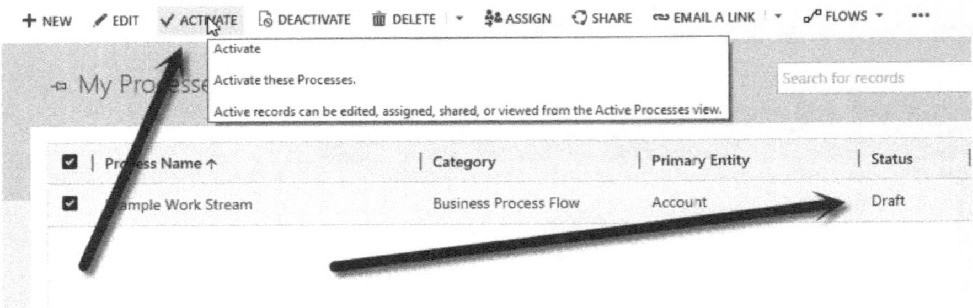

Figure 6.13: Activating the business process flow.

The activation process (Figure 6.14) can take a while to complete.

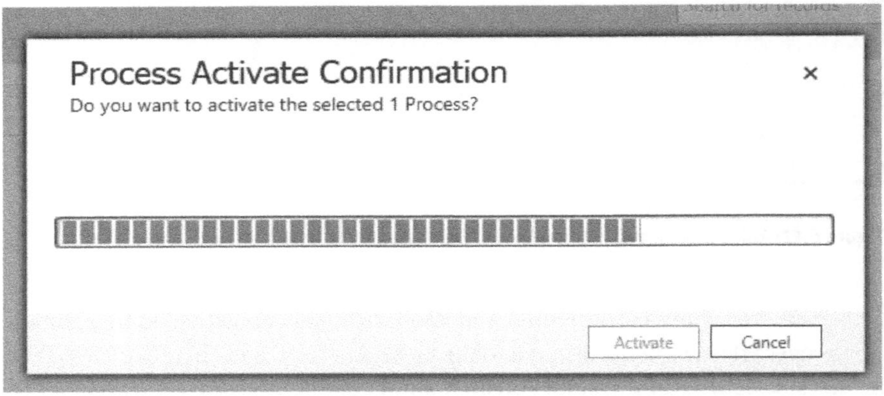

Figure 6.14: The activation process can look like it is stuck at the activation step—be patient.

With the process activated, you can now browse to your entity where you will see the BPF ribbon at the top of the window. In the case of the example we just looked at, open up an account record. You will see the ribbon shown in Figure 6.15. You may need to click on the Process button and start your BPF in order to see this the first time.

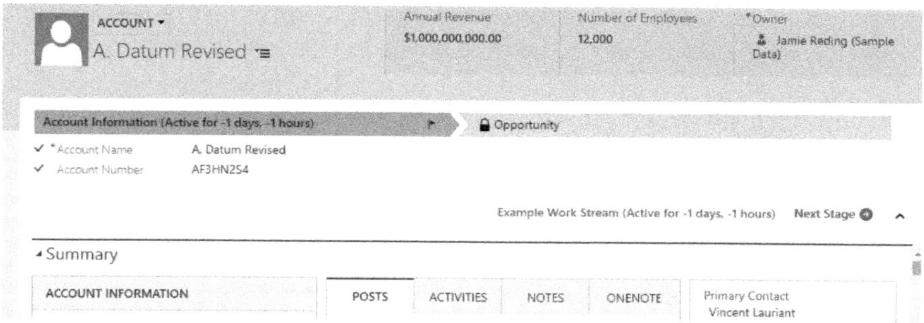

Figure 6.15: A new ribbon will appear on your form once the BPF has been activated.

You have successfully deployed a very simple business process flow based on the steps in this section. Typically, a BPF will have multiple stages and many fields rolled up to the header. It will draw from multiple entities, and be used to assist the user in getting the appropriate information filled in at each step of your business's human workflow stages. In this example, we've linked it to an Opportunity. When you click Next Stage (Figure 6.16), you will be able to link this account to a new or existing opportunity record.

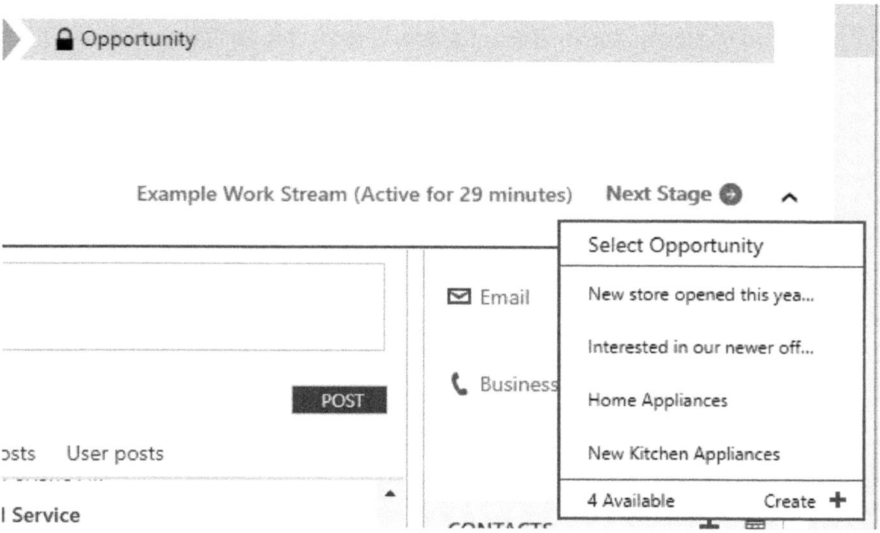

Figure 6.16: Moving on to the next stage in the BPF of the entity.

Implementing Business Process Flows allows you to truly capture the power of Dynamics 365. Users no longer have to navigate through complex forms and across entity forms. Instead, they can work through a process from beginning to end within a single view. Learning to leverage BPFs throughout your implementation can greatly improve efficiencies for your user base.

Business Rules

Business rules can be created and applied to form logic without writing JavaScript code or creating plugins. Business rules provide a simple interface to implement and maintain fast-changing and commonly used rules. They can be applied to Main and Quick Create forms. Business Rules are limited to the following actions:
- Show error message
- Set field value
- Set Business Required
- Set Visibility
- Set default value
- Lock or unlock a field

To create a business rule, take the following steps:
1. Navigate to a specific entity within your solution (via the Customizations menu) and click on the Business Rules tab, as shown in Figure 6.17.

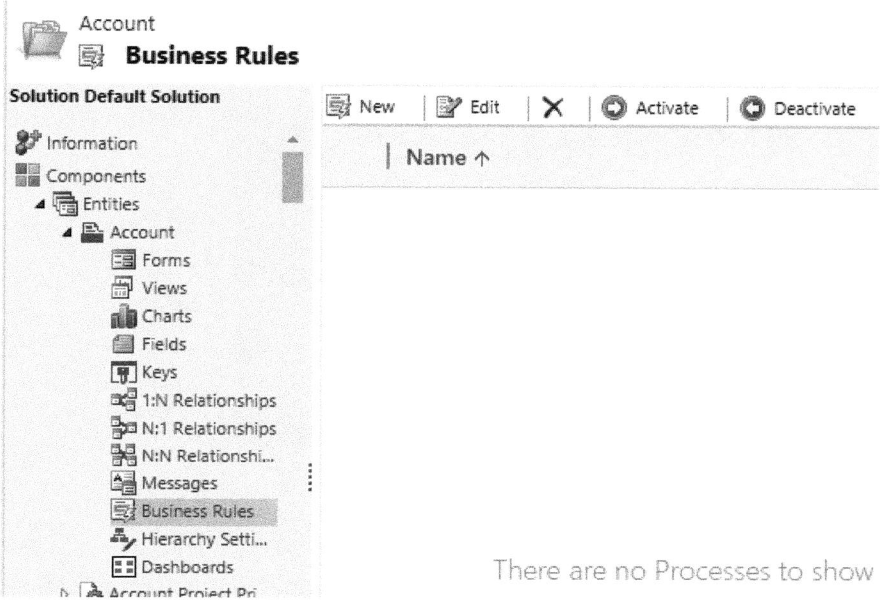

Figure 6.17: Accessing the business rules from the solution.

2. From here you will be able to click the New button. A development surface that looks very similar to the Business Process Flow surface will open (Figure 6.18). This is where you will be able to drop and configure tiles that represent the steps within your business rule.

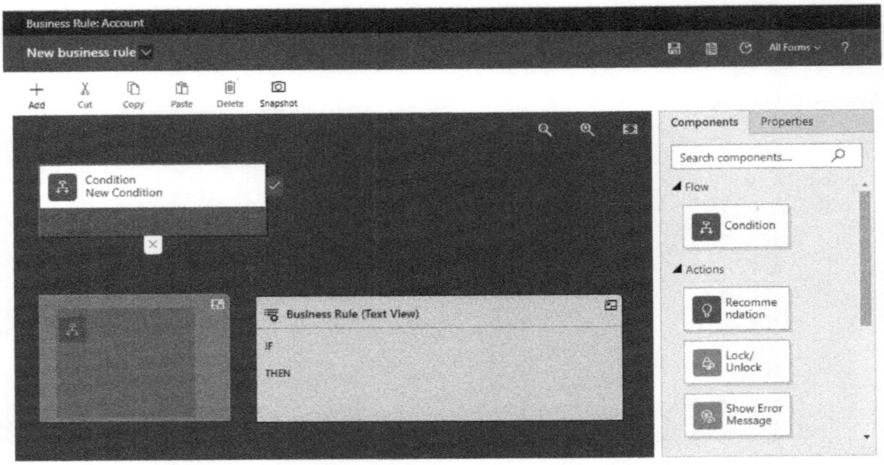

Figure 6.18: The design surface for the BPF.

3. Click on the Condition tile that is automatically created. We will create a simple rule that states that the Customer Size field on the account must have a value. First, set the Display Name to something descriptive, then define the rule. In this case, set the Source to Entity, the Field to Customer Size, and the Operator to Does not contain data. Click Apply when complete and the Figure 6.19 statement will show. This will create the "IF" portion of the business rule.

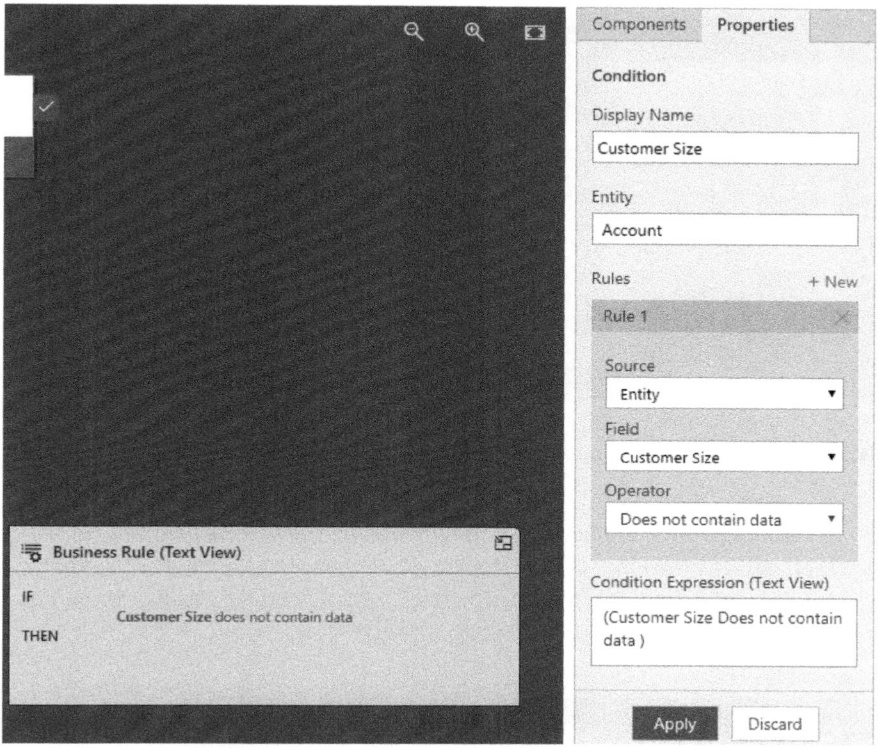

Figure 6.19: Creating the initial condition.

4. Now, click on the Components tab. You will need to drop a tile that represents what you want to have happen when the condition is true. In this case, we'll just show an error message, so drag a Show Error Message tile out onto the design surface and drop it on the plus sign that appears to the right of the Customer Size Condition you just created.
5. On the Properties tab, set the information for this Error Message. The Field property will need to be pointed back at the Customer Size field. The Message property is a free text box where you can type whatever message is appropriate. Once done, click Apply. Figure 6.20 shows the fully configured condition.

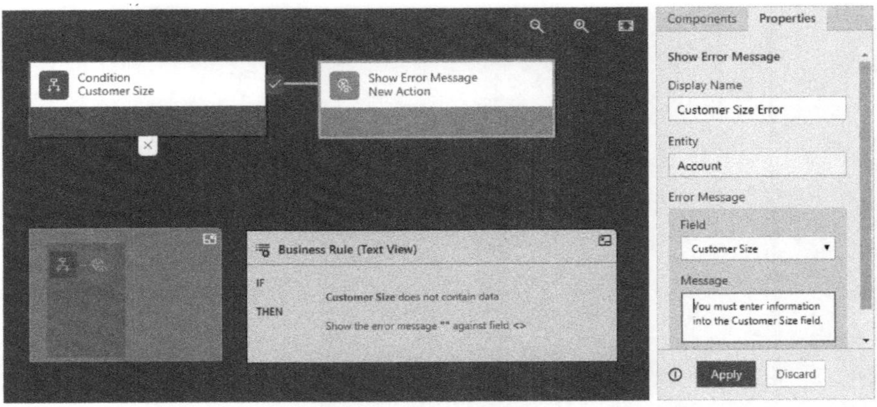

Figure 6.20: Adding an error message as the result when the condition is met.

6. When done, click the Save button followed by the Validate button (to see if there are any errors). You can then activate the rule by clicking the icon that has a greater than sign inside of a circle, which is on the toolbar to the right of the Save button (Figure 6.21).

Figure 6.21: The Activate button is hidden on the toolbar.

In the older versions of Microsoft Dynamics CRM, a developer would have to write a lot of JavaScript to handle standard business rules. For example, if you had a field that you wanted to disable from being edited when another field had a certain value in it, you'd have to write JavaScript to disable that field. This type of functionality can now be handled through a configured business rule. The other standard actions that you can implement in a business rule are shown in Figure 6.22.

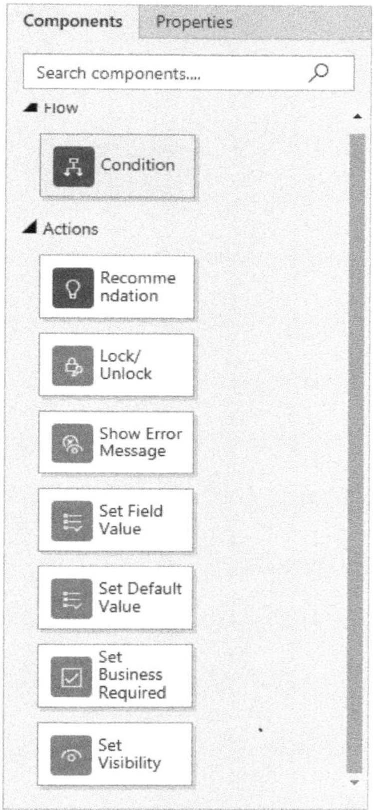

Figure 6.22: There are many options available for a business rule.

Summary

The proper use of business process flows and business rules in Dynamics 365 can transform the user experience, and create a platform that can greatly improve overall efficiencies. In most CRM platforms, users must know what to click on and how to navigate across forms. In Dynamics, with business process flows in place, a user can quickly create and update information within a single form without having to spend time clicking through the system to find data. Business rules can greatly reduce development time (there is now limited need for JavaScript to alter form functionality), and can direct users to enter information in a more logical way.

Chapter 7
Workflows and Other Processes

Dynamics 365 has robust capabilities for building automated workflow processes that perform actions when events trigger or when users manually initiate them. A wide array of actions can be performed, including creating and updating records, sending emails, calling external systems to insert or retrieve data, querying other entities within CRM, interacting with the user to get information, and performing complex routines in code. This chapter will outline how to create and monitor workflows, and will introduce two other process types called Dialogs and Actions.

Accessing Workflows

There are three ways you can navigate to workflows from within your Dynamics 365 environment:

1. You can open Customizations from the main Settings menu and click on Customize the System. In the Default Solution window that opens, you can look in the Processes tab (see Figure 7.1).

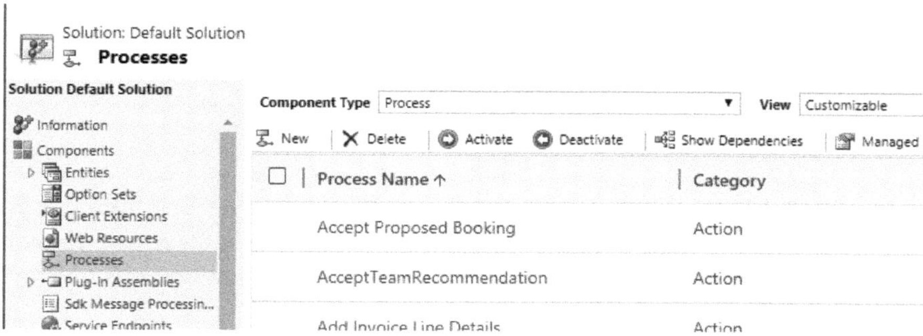

Figure 7.1: Processes can be accessed from a solution.

2. You can search for the Process entity from within the Advanced Find window. Note that there will be a number of preset filters when you do this (as shown in Figure 7.2). You will need to keep these filters. If you ever want to search for Processes that relate to a specific entity, you can click on the Primary Entity attribute.

DOI 10.1515/9781547400065-007

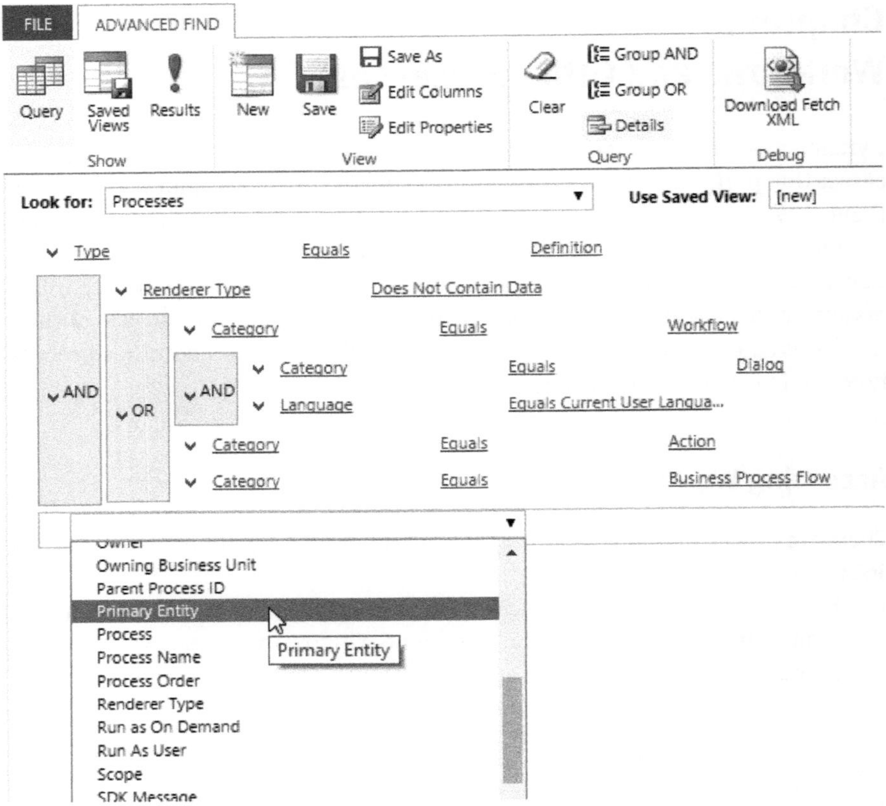

Figure 7.2: You can use Advanced Find to access processes.

3. As the third option, you can navigate to Settings and then select Processes under the Process Center (Figure 7.3). This will take you to the classic view of your workflows, and allow you access to the various views that you get on any other entity.

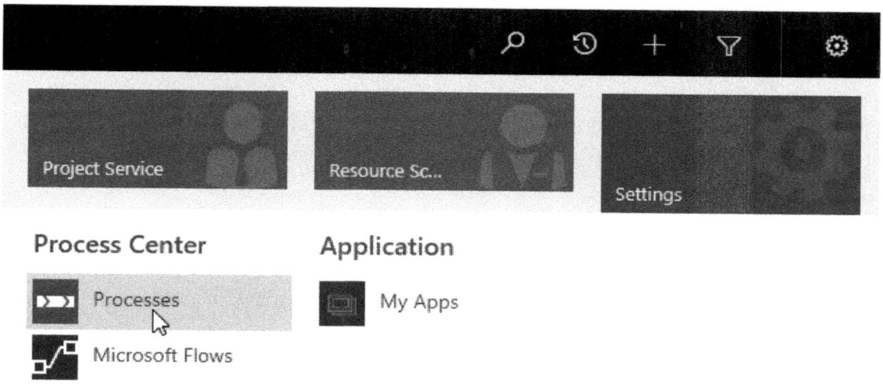

Figure 7.3: You can use the main menu to navigate to Processes.

From any of these three areas, you can edit existing workflows or create a new workflow. Note that in these views, you will see not only workflows, but also business rules and other types of components. You will see the Workflow type specified under the Category heading, as shown in Figure 7.4.

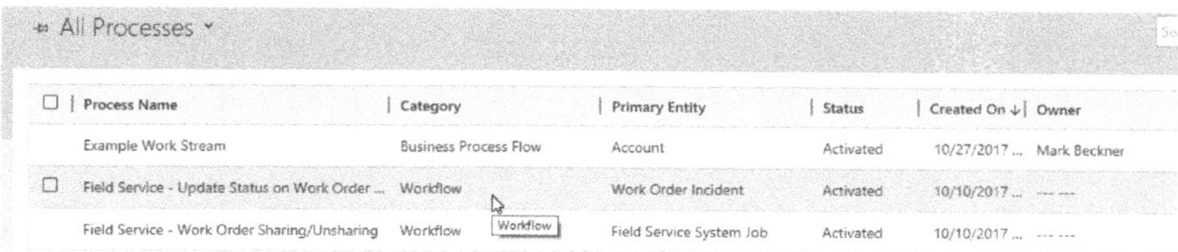

Figure 7.4: The Category column will indicate the type of process you are looking at.

Creating a Basic Workflow

To demonstrate how to create a workflow, we'll work through a very simple example. In this example, we will send an email when a new account is created.

To create a workflow, click on New from within any of the three areas that you have navigated to. Clicking this button will open a dialog box where you can set the name of the workflow, the entity that it is attached to (in this case, Account), and whether it will be entirely new or based on an existing template. You must also select Workflow as the value in the Category property. Finally, you will see an option called “Run this workflow in the background.” This should be kept checked, and allows the workflow to run asynchronously (this means that the user can continue to do work while this process runs independently). An example configuration is shown in Figure 7.5.

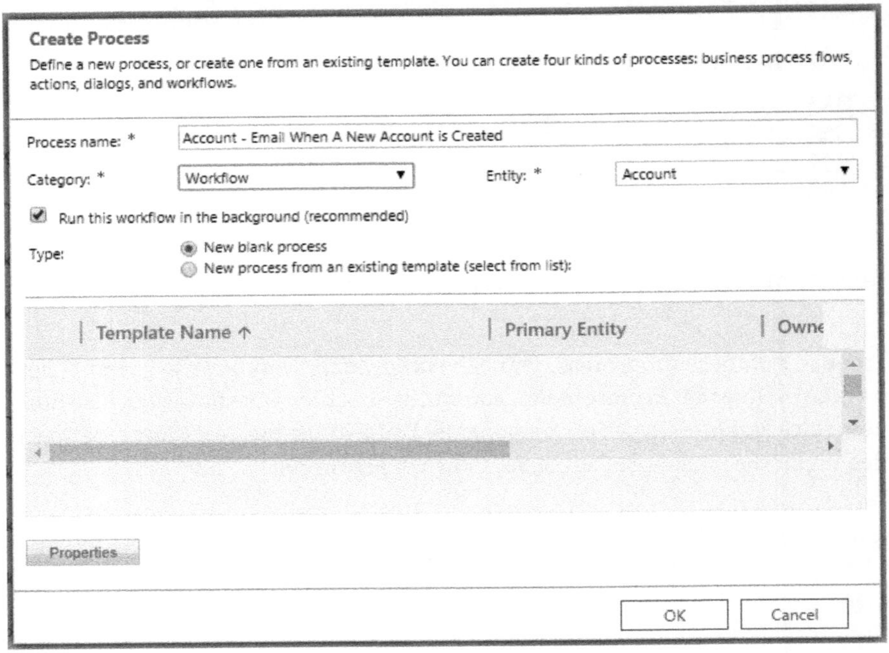

Figure 7.5: Creating a new workflow process.

Once you have the properties in the box set, click the OK button. This will open a new window where you can build out the steps in your workflow. There are several things to note on this new screen (which is shown in Figure 7.6):

- The option “As an on-demand process,” if checked, will allow users to initiate this workflow manually from within a specific entity. For example, if you had a workflow that did some sort of validation logic on a record, you could allow a user to manually kick this off. In the case of the example we are looking at, which is the automatic sending of an email when an account is created, we do not want to set the workflow as an on-demand process.
- The “As a child process” option allow you to create a workflow that can be called from another workflow.

- You can set the Scope property to User, Business Unit, Organization, or Parent: Child Business Units. These options let you determine who can access the workflow. You can create a workflow that is only accessible to specific users or business units, or you can create a workflow that is accessible to everyone in the organization. In the case of an automatic email being sent on account create, select Organization, since you will want this available to be run by the full organization and not be connected to the actions of a specific user.
- The "Start when" bulleted list lets you determine what action on the entity causes the workflow to trigger. In the case of this example, it will be set to "Record is created."

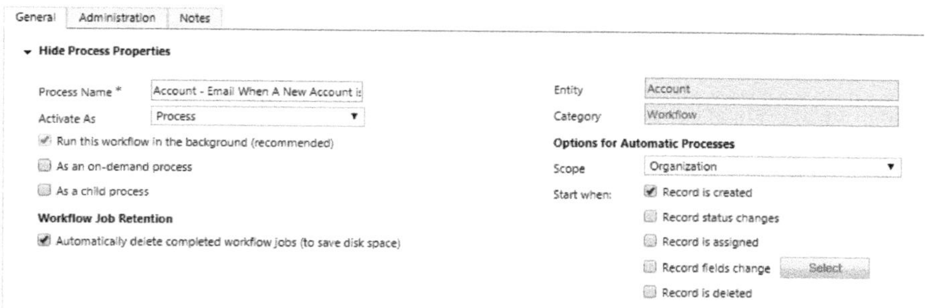

Figure 7.6: Configuring the workflow.

At the bottom of this window is where you will be able to configure the steps within your workflow. There are many options within the Add Step drop-down menu, and you can create workflows with heavy logic and many steps. More information on adding logic to your workflow will be covered later in this chapter. For this example, click on the Send Email option, as shown in Figure 7.7.

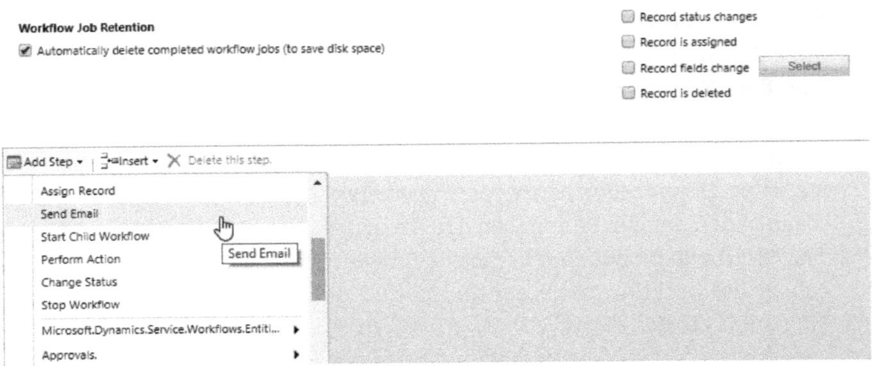

Figure 7.7: Adding a Send Email step to your workflow.

Selecting Send Email will add the step to your workflow surface. You need to give the step a descriptive name and then click the Set Properties button (see Figure 7.8). As you name the step, give as much detail to the description as you can. In cases where you have really involved logic, you, and others who are supporting your Dynamics 365 implementation, will need to read through the workflow steps and be able to determine what is going on without having to open individual steps for review.

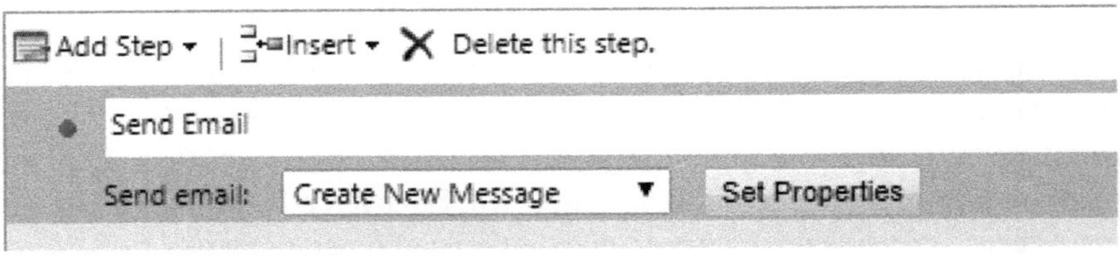

Figure 7.8: Once a step is added, you will need to click Set Properties.

When you click the Set Properties button, a new window will open where you'll be able to configure the information you need to send the email and populate the text of the email. You can set standard information like the To, From, Cc, and Subject in the header. Setting the To and From is fairly simple as long as they are static addresses (meaning, it is the same To and From every time an email is sent). Click the lookup button on the To and From fields. This will open a Look Up Records dialog box (see Figure 7.9). Click next to the account that you want to target, and then click the Select button. In this example, in order to send emails to the applicable internal personnel, you will want to query for Users, and select the applicable user accounts set up in your system. Once the users you want are shown in the Selected records field, you can click the Add button and the user or users will appear in the To or From field you just set.

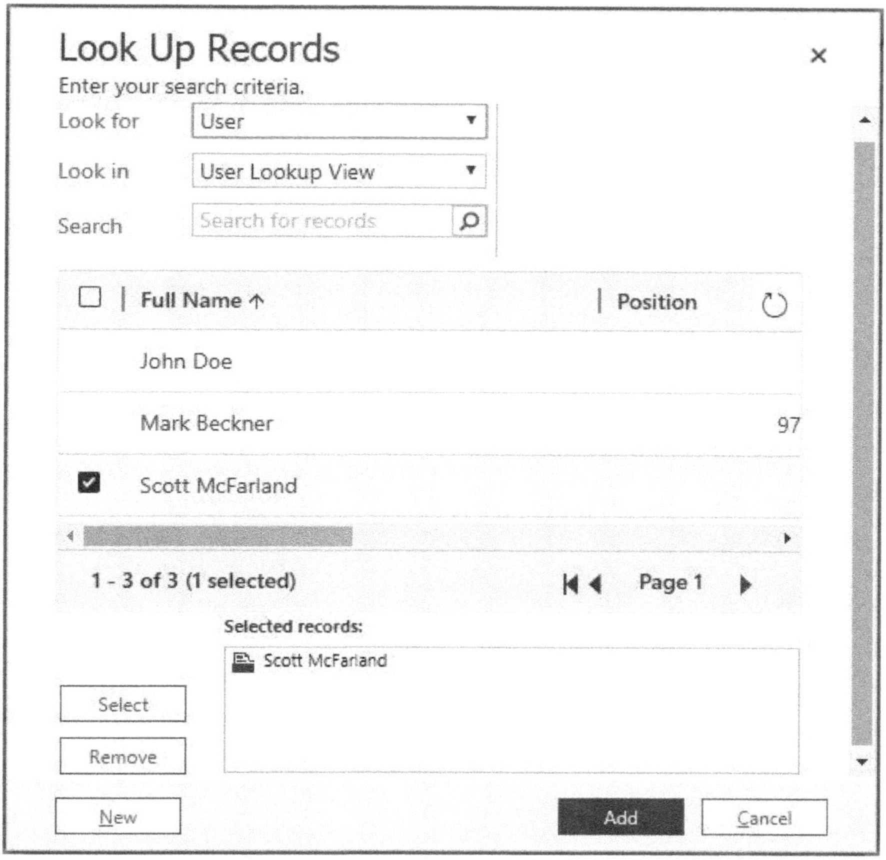

Figure 7.9: Looking up the information for the To and From fields in the email configuration.

In most cases you will want to send dynamic information in your emails. For example, let's say that you wanted your subject line to display the account name of the account just created. Click in the Subject field of the email configuration form and then look on the right side of your window. You will see an area called the Form Assistant where you can select what field you want to include. In this case, you'll want to set it to look for the Account Name field in the Account record you are on. Refer to Figure 7.10 for how this is configured.

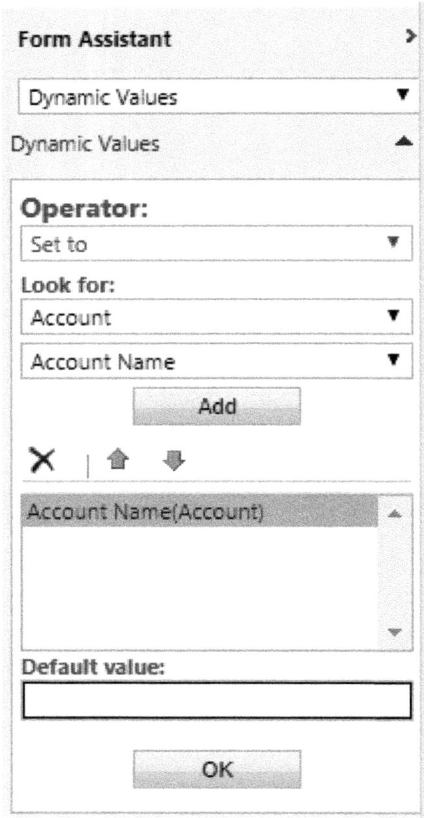

Figure 7.10: The Form Assistant is where you can select fields for dynamic information in your step configuration.

Click OK once you have added the Account Name field to the Form Assistant. This will add the field to the Subject property, and highlight it in yellow. A dynamic field will generally be highlighted in yellow within your configuration screen, but not always. The way to know that a field is dynamic is to look at any text wrapped in {curly brackets}. Figure 7.11 shows the Form Assistant being used to populate a dynamic value in the Cc field. This value comes from the related user record.

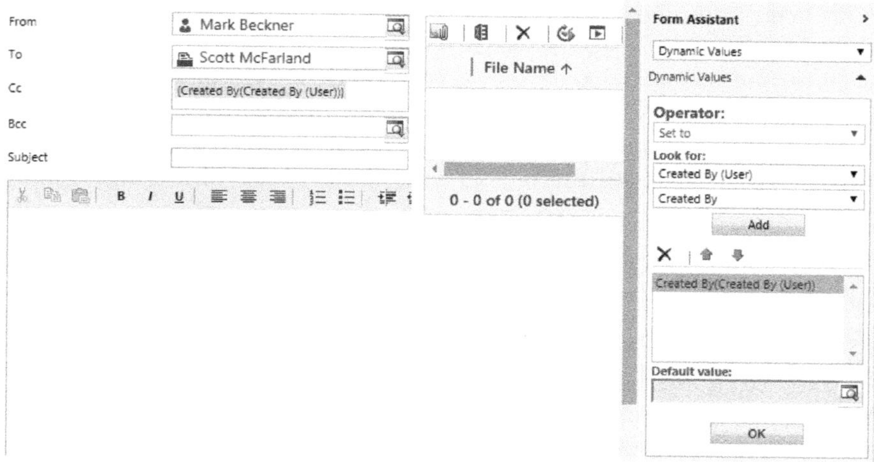

Figure 7.11: The Form Assistant is where you can select fields for dynamic information in your step configuration.

Figure 7.12 shows the subject line with the dynamic Account Name added to it, as well as dynamic fields in the email body.

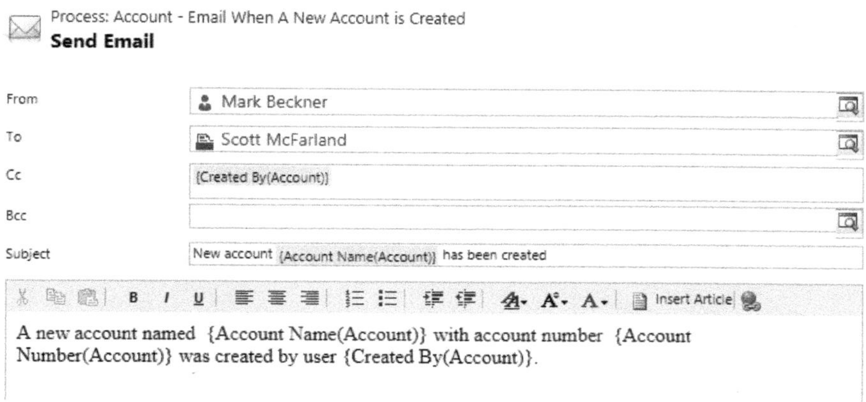

Figure 7.12: Dynamic information being added in various fields of the email.

To add dynamic information in the body of the email, first type the email. Then place the cursor at the point you want the dynamic field to be entered, set the Look for values, click Add, then click OK. Once the dynamic field has been added to the body of the email, you can drag and drop the fields to place them where you want.

When you are done configuring the email, click the Save and Close button to complete the step. This will return you to the main workflow configuration screen. You can now save this workflow by clicking the Save button and then activate it by clicking the Activate button. Once activated, you will no longer be able to edit the workflow. If you want to make edits, you'll have to click the Deactivate button.

Monitoring Workflow Status

At this point, with the process activated, you can create an account that will cause this workflow to fire and an email to be sent. When a workflow fires, you'll be able to monitor the status and progress, as well as any errors that may have occurred. In order to monitor the current email example, create an account (make sure your workflow is activated first).

There are several ways to see the workflow process status. The first is to click on the Process Sessions tab within the workflow configuration screen you have already been working within. In Figure 7.13, you will see that there has been one instance that has been kicked off of this process.

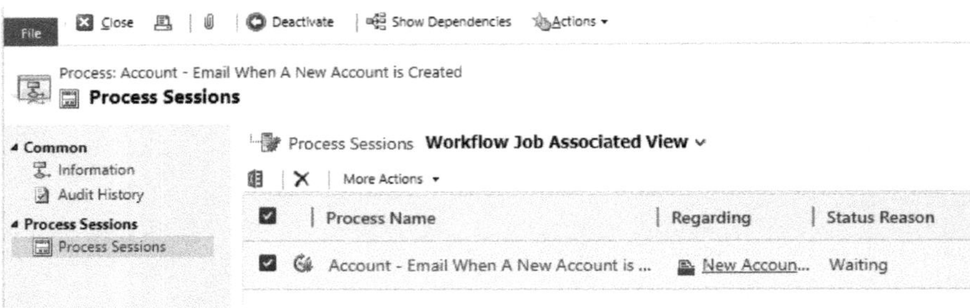

Figure 7.13: Looking at the Process Sessions tab.

The Status Reason is Waiting, which means the workflow has been initiated, but cannot complete. Click this instance and details about what has occurred will be shown. As you can see in Figure 7.14, there is an issue with the email addresses on at least one of the accounts. Clicking on the Details tab will give you the system exception generated which should allow you to troubleshoot this issue. In this example, the account that was selected as the recipient did not have a valid email address associated with it, so the email could not be sent.

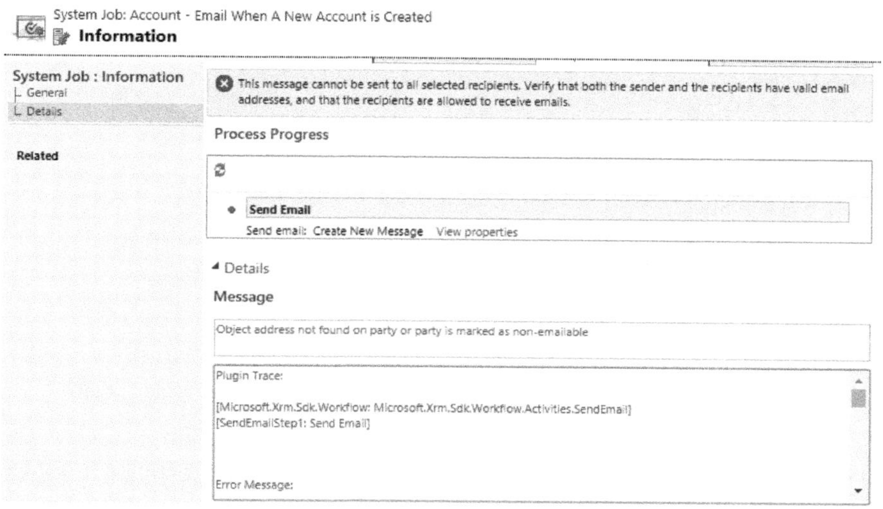

Figure 7.14: You can see the error that has occurred on a failed workflow.

Another way to access the status of a workflow is to click on System Jobs from the main Settings menu. This will give you a list of all processes that have been triggered, and will allow you to assess health across your system in a single view. Processes shown here include custom workflows and system workflows.

As an administrator of Dynamics 365, you will want to check the System Jobs page periodically to view the health of your system and address any issues that may be occurring with your workflows.

On-Demand Workflows

If you check the "As an on-demand process" option on the General tab of the Workflow (Figure 7.15), you can allow your workflow to be initiated by a user directly from a specific record.

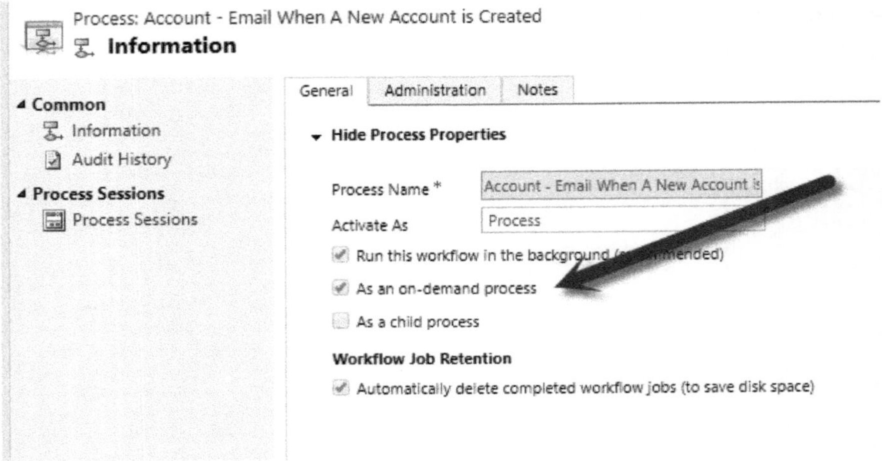

Figure 7.15: Allow users to manually start workflows by selecting the on-demand option.

In the case of the example workflow we just created, if you Deactivate it, select the on-demand option, and Activate it again, you will be able to manually initiate it from an account. Once you activate it, navigate to a list of account records and select one of the records. Then, click the ellipses in your toolbar and select Run Workflow, as shown in Figure 7.16.

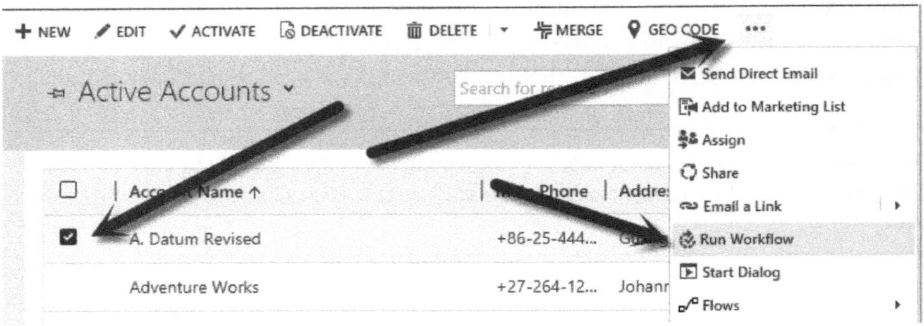

Figure 7.16: A user can select a record and then click the Run Workflow option from the menu.

After you click Run Workflow, you will be prompted to select which workflow you want to initiate. When you select the workflow, a dialog box will force you to confirm whether you want to continue or not. If you click OK on the dialog box shown in Figure 7.17, the workflow will be scheduled to run. It can take a while before a workflow will begin processing. You can check the status of the workflow as indicated earlier to see if it is executing as expected.

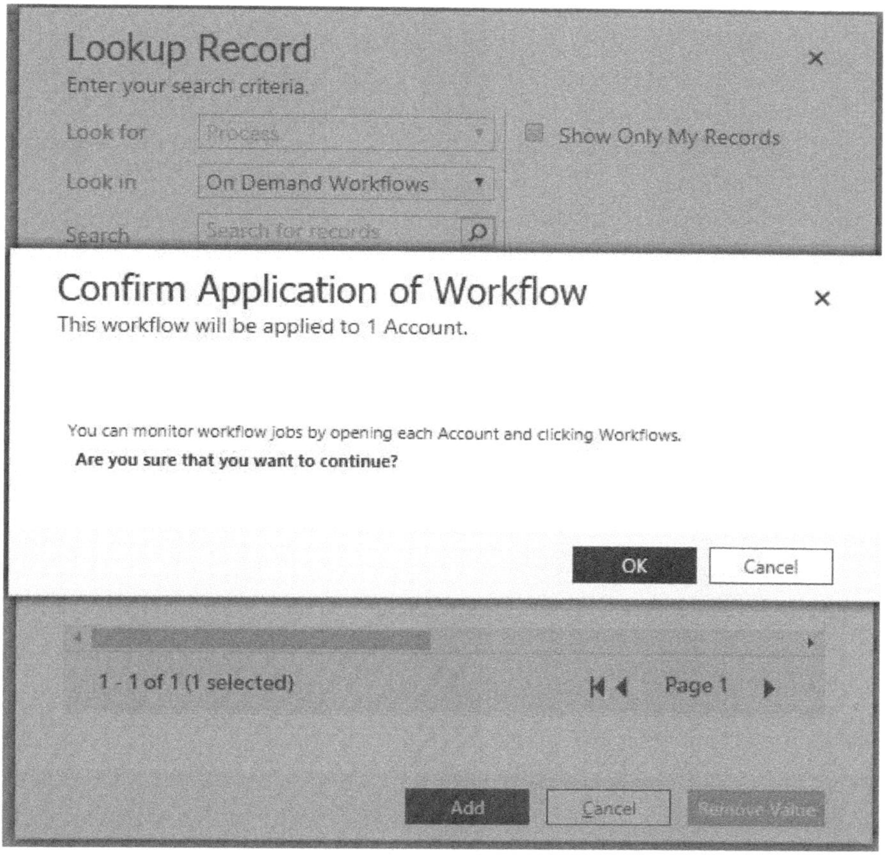

Figure 7.17: Confirmation provided for the user to run the workflow.

Adding Logic to a Workflow

Sending an email on creation of a record is a very simple application of a workflow. You can create any level of complexity within workflows that you need, and it is common to have multiple steps and conditional logic. As you build out processes in workflows, think about the following:

1. Workflows give business analysts the ability to create logical flows without having to reach out to developers. The challenge is to know how much complexity is too much. Workflows with a lot of logic in them can become unwieldy to manage and test.
2. When you are developing a workflow, take care in architecting it accurately, and ensuring that you have tested all cases.
3. Determine what it will take to make modifications to your workflow once it has been deployed in a production setting. If others are going to have to modify it, will they be able to easily understand what has been configured?

There are many logical options available to you. For example, let's say that you only wanted to send an email on account creation when the account annual revenue was in excess of $100,000. The first thing to do would be to add a Check Condition step to your workflow (see Figure 7.18).

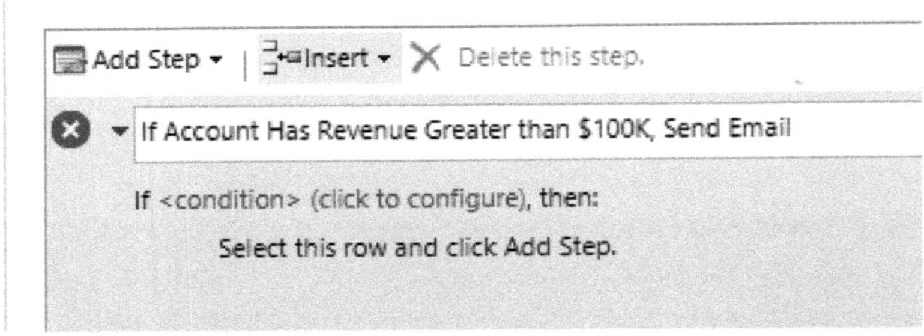

Figure 7.18: Adding a check condition step.

The next thing to do is click on the condition link (it will say "<condition> (click to configure)" and work with the condition logic window that opens. Figure 7.19 shows how to configure for the annual revenue example.

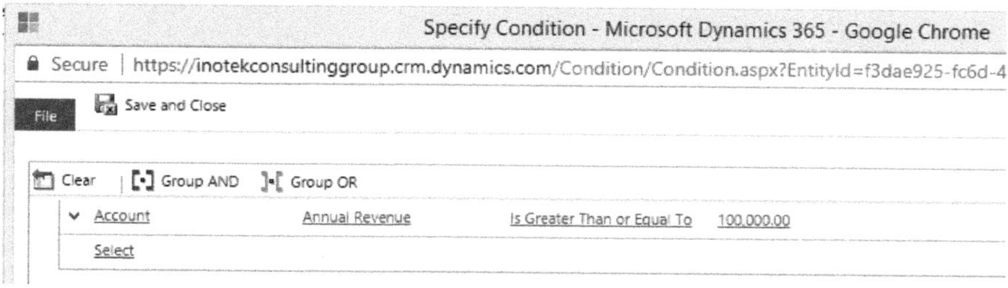

Figure 7.19: Specifying the logic of the condition.

Then click Save and Close and return to the main workflow configuration screen. Under the condition that you just configured, add a new step to send an email (like that shown in Figure 7.20). You can configure this email send the same way as you did in the earlier example from this chapter. Notice that you don't have to configure any other logic (like an "Else") in this scenario. If the workflow executes and the condition is met, the email will fire, otherwise the process will simply end.

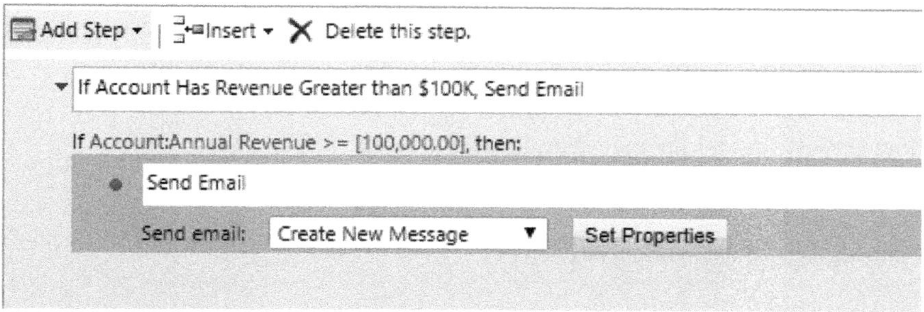

Figure 7.20: The full condition has been configured.

This is an example of fairly simple logic. You can build workflows that get extremely involved, like the one shown in Figure 7.21.

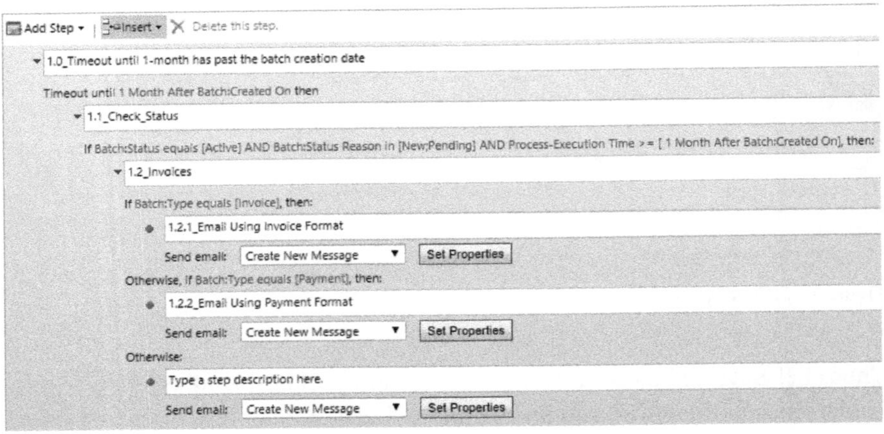

Figure 7.21: An example of nested logic in a workflow.

When you go to configure a workflow, first become familiar with all of the options available to you within the workflow configuration window. Once you understand what is there, you can determine the best way to build out your workflow. If you find that it is something that will require a lot of logic or may require heavily nested conditional statements, you may decide that it is something more easily and cleanly (or efficiently) done in a custom workflow activity and C# code. There are many cases where the only step in a workflow is to call a custom workflow activity and let the code handle all the work.

Creating Custom Workflow Activities

There are times when the logic of a workflow is so involved that it is best done through code. You may have the need to call web services that live in third party systems outside of Dynamics, you may have lookups that need to occur across many entities within your data model, or you may have complex calculations that need to occur. In whatever case, you can create what is called a Workflow Activity in C# and call this C# activity from within a workflow.

Workflow activities are very similar to plugins. They are coded in a similar fashion and deployed in the same manner. The key difference is that a workflow activity can only be called from within a workflow, whereas a plugin can only be triggered off of an action on an entity. A workflow activity can also have parameters set to be passed into it from the calling workflow.

Listing 7.1 shows the C# framework you would need to build in order to code a workflow activity, along with some example code to show how functional code would be implemented. Earlier you looked at sending an email as a step within a workflow. This example code also shows how to send an email. This will give you an idea of the work involved in coding something from scratch.

Please refer to the chapter on Plugins in this book to see more detail around building and deploying code within Dynamics 365.

Listing 7.1. Sample Workflow Activity Framework

```
// the CRM SDK will need to be referenced in your
development
using Microsoft.Crm.Sdk.Messages;
using Microsoft.Xrm.Client;
using Microsoft.Xrm.Portal;
using Microsoft.Xrm.Sdk;
using Microsoft.Xrm.Sdk.Query;
using Microsoft.Xrm.Sdk.Discovery;
using Microsoft.Xrm.Sdk.Messages;
using Microsoft.Xrm.Sdk.Client;
using Microsoft.Xrm.Sdk.Workflow;
using Xrm;
namespace Inotek.Examples
{
 public class SendEmailExample : CodeActivity //
CodeActivity must be inherited
 {
 // here is how you set input parameters that can be passed
in from your workflow
 [RequiredArgument]
 [Input("FROM party for generated Email")]
 [ReferenceTarget("systemuser")]
 [Default("9D61E95F-8F2B-E311-A85C-0050569723F2",
"systemuser")]
 public InArgument<EntityReference> FromUser { get; set; }
 // second parameter
 [RequiredArgument]
 [Input("TO party for generated Email")]
 [ReferenceTarget("contact")]
 public InArgument<EntityReference> ToUser { get; set; }
 /// <summary>
 /// Execute method that creates the email
 /// </summary>
 protected override void Execute(CodeActivityContext
executionContext)
```

```
{
//Create the tracing service
ITracingService tracingService =
executionContext.GetExtension<ITracingService>();
//Create the context
IWorkflowContext exContext =
executionContext.GetExtension<IWorkflowContext>();
IOrganizationServiceFactory serviceFactory =

executionContext.GetExtension<IOrganizationServiceFactory>()
;

IOrganizationService service =
serviceFactory.CreateOrganizationService(exContext.UserId);
EntityReference _userId;
EntityReference _toContactId;
_userId = FromUser.Get(executionContext);
_toContactId = ToUser.Get(executionContext);
Guid fromUserId = FromUser.Get(executionContext).Id;
Guid toContactId = ToUser.Get(executionContext).Id;
// Create the 'From:' activity party for the email
ActivityParty fromParty = new ActivityParty
{
PartyId = new EntityReference(SystemUser.EntityLogicalName,
fromUserId)
};
// Create the 'To:' activity party for the email
ActivityParty toParty = new ActivityParty
{
PartyId = new EntityReference(Contact.EntityLogicalName,
toContactId)
};
// Email body
Email email = new Email
{
To = new ActivityParty[] { toParty },
From = new ActivityParty[] { fromParty },
Subject = "Example Email from Workflow Activity",
Description = "[Call another method here to come up with
dynamic content]"
};
Guid _emailId = service.Create(email);
// Send the email
SendEmailRequest sendEmailreq = new SendEmailRequest
{
EmailId = _emailId,
TrackingToken = "",
IssueSend = true
};
```

```
 SendEmailResponse sendEmailresp =
 (SendEmailResponse)service.Execute(sendEmailreq);
 } // End Execute
 } // End Class
} // End Namespace
```

When you deploy a Workflow Activity, it will appear in the list of available steps in your workflow configuration screen. In Figure 7.22, the arrow shows workflow activities that are available out of the box with your Dynamics Online instance. When you deploy your own workflow activity, it will appear in this area of your steps. In order to add it as a step in your workflow, click the arrow to the right and select the method that you want to call.

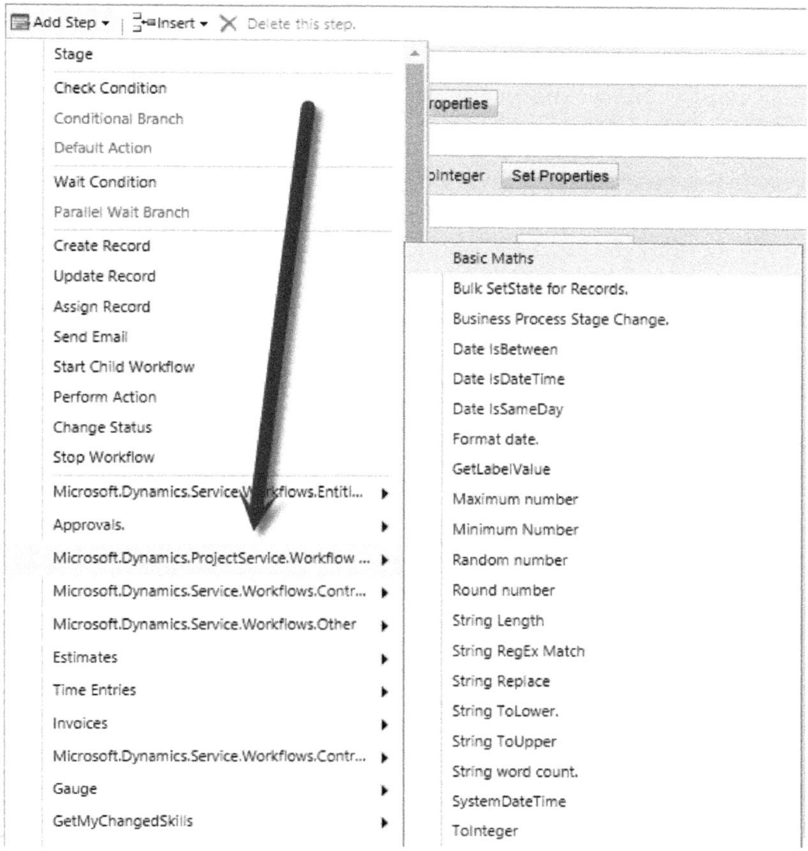

Figure 7.22: Your workflow activity will appear in this section of the Add Step options.

You can see in the preceding case that there are many math functions available in the workflow activity that has been selected. When Basic Maths has been selected, it will be added as a step in your workflow, as shown in Figure 7.23.

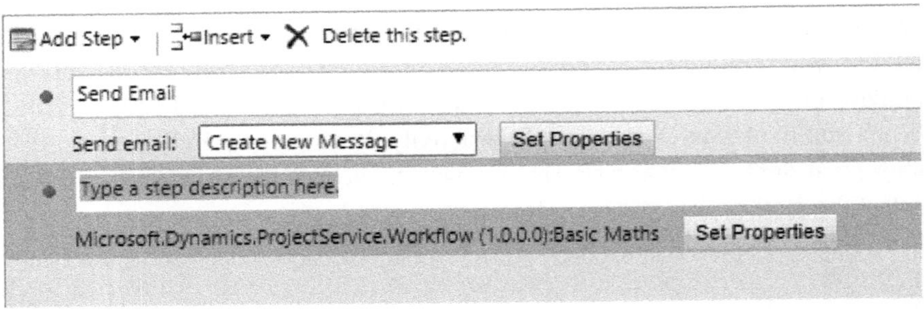

Figure 7.23: Selecting the Basic Maths workflow activity.

If your workflow activity has parameters that need to be passed in, you will be able to configure them when you click the Set Properties button. Figure 7.24 shows that the Basic Maths method has several fields that need to be set. In this case, we will multiply the first number (of .055) by the second number (which is a dynamic value from the account entity's Account Revenue field). When you have configured the input parameters for your workflow activity, you can click Save and Close to complete the process.

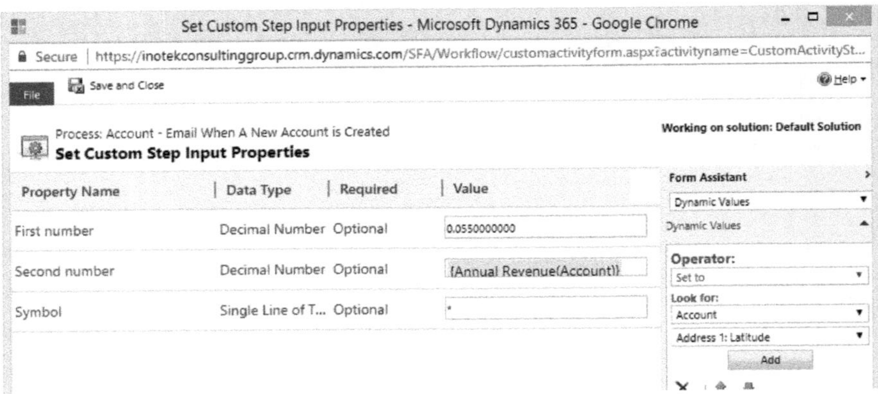

Figure 7.24: Configuring the inputs on a workflow activity.

Dialogs

Another type of process that is similar to workflows is the *Dialog* process. This allows interaction between the user and CRM to influence the logic within the workflow or to add real-time data for processing. An example of this is shown in Figure 7.25, where the dialog is called in order to prompt the user for credit card information. This can be called directly from a workflow, if needed. To create a dialog, take the following steps:

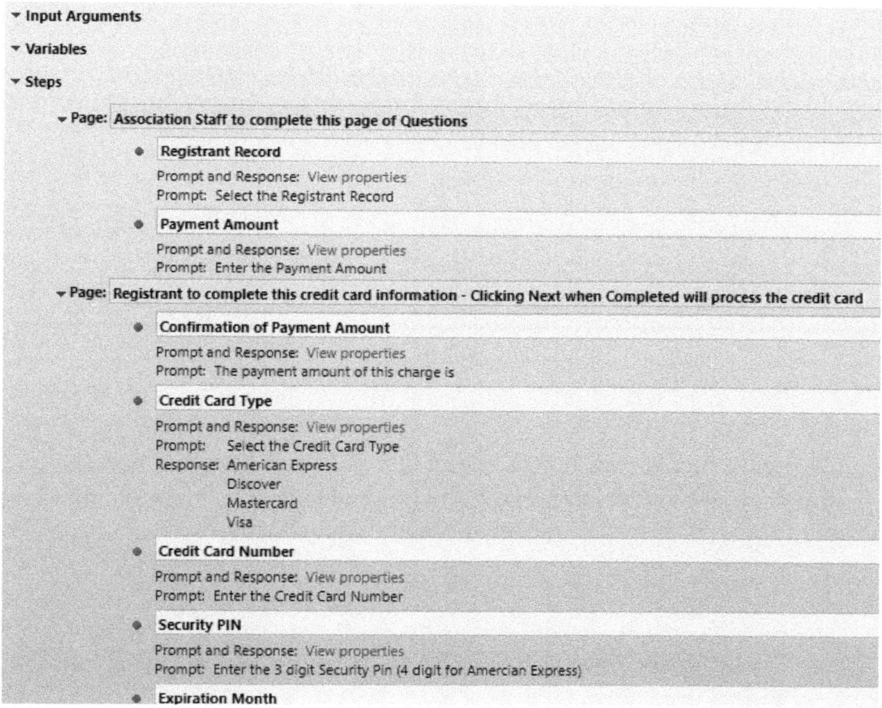

Figure 7.25: An example of a credit card dialogue process.

1. From a solution click on the New button. The Create Process screen shown in Figure 7.26 will open. Enter a process name, select the entity this dialog will be associated with (in this example, Account), and choose Dialog in the Category drop-down.

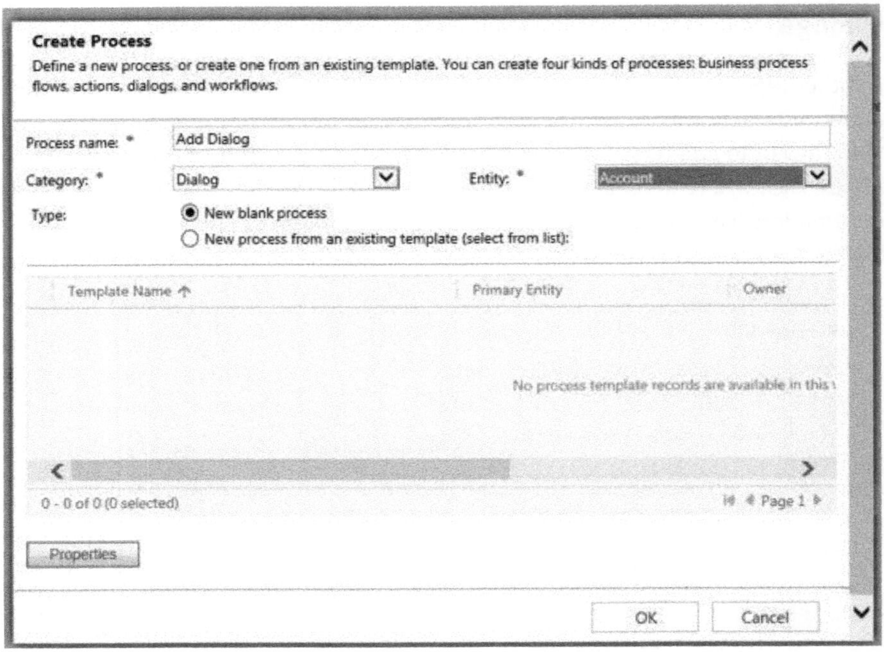

Figure 7.26: Creating the dialog.

2. When you click the OK button on the Create Process screen, a new window will open (see Figure 7.27), which will be familiar to you. It has a similar layout to other process types covered earlier in this chapter.

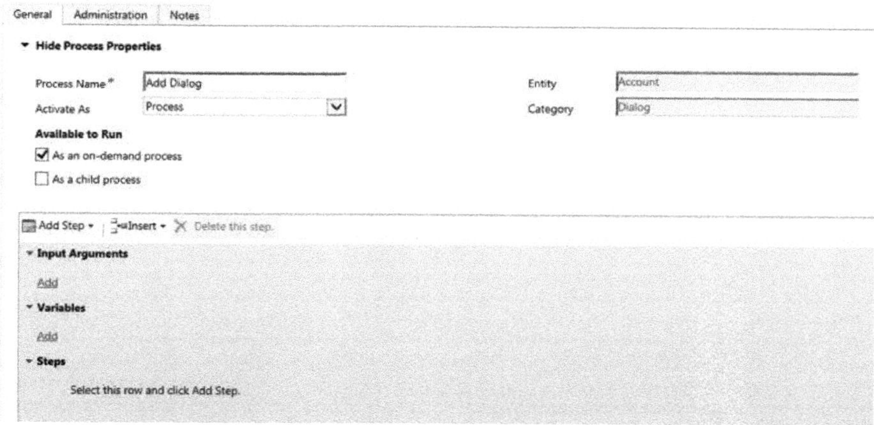

Figure 7.27: Configuring the dialog.

3. Select the "As an on-demand process" if you want a user to be able to manually start the dialog. Set the "As a child process" flag if you want to be able to call the dialog from a workflow or other process. For this example, select the "As a child process" check box.
4. Set the Input Arguments. This can only be done if the dialog is a child of another process. An on-demand dialog cannot have input parameters. Figure 7.28 shows a dialog that is set up as a child process which has four input parameters of various data types. As you can see, default values can be set if needed.

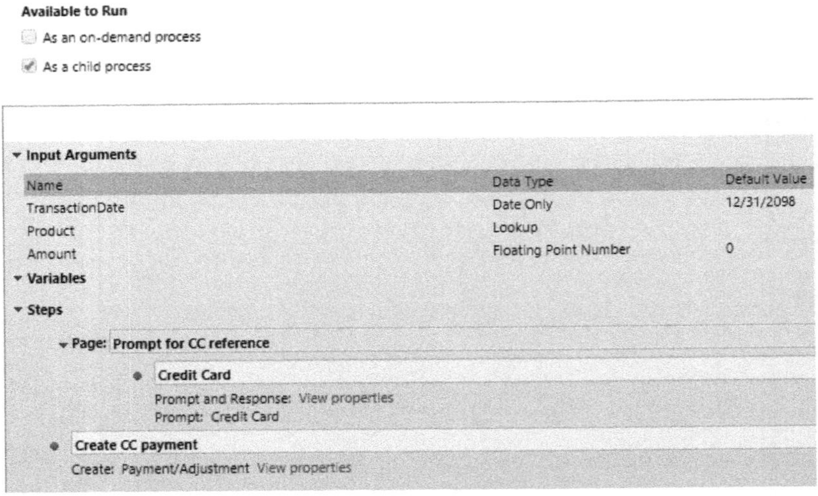

Figure 7.28: Input parameters can be defined when the dialog will be a child of another process.

5. The Variables section allows for the configuration of variables related to this dialog. Variables can be of several types, including a single line of text, whole numbers, floating point numbers, date/time, date only, and lookup. Variables can be referenced in the Steps of the process flow.
6. In the Steps Area, you will define the process flow for what the dialog will do. You can select from a variety of options, but most importantly you can select the Page option (see Figure 7.29).

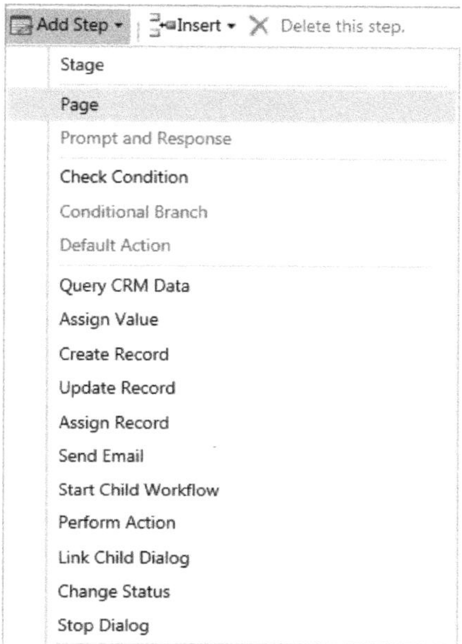

Figure 7.29: Select the Page option to allow for Prompt and Response with users.

7. When selecting Page, the Prompt and Response screen in Figure 7.30 will open. On this page you can:
   a. Enter a Statement Label.
   b. Enter the text to be read in the Prompt Text area.
   c. If any tips are to be provided to the user, these can be entered in the Tip Text area.
   d. The bottom section of the form will be enabled or disabled based on the option selected in the Response Type. Response Types include Single Line, Option Set, Multiple Lines, Date and Time, Date Only, and Lookup. Selecting one of the Option Set types will enable the lower section. You can set hardcoded values, or you can do lookups against CRM for dynamic values.
8. When you are done with the configuration, click the Save and Close button and then activate your dialog to complete the process.

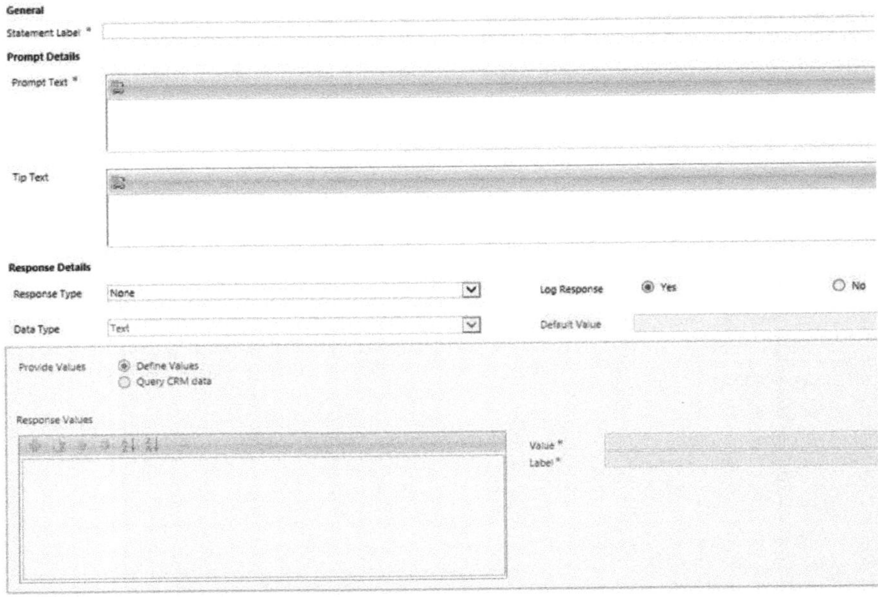

Figure 7.30: Configuring the Prompt and Response information of a dialog.

Actions

There is yet another type of process available in Dynamics 365 that is similar to a workflow, except that it can be called directly from a variety of locations, including plugins, workflows, JavaScript within a form, and other code. *Actions* are configured in a very similar way to workflows. In order to create an action, you can select the "Action" option from the Category drop down when creating a new Process (see Figure 7.31).

Actions are different from dialogs in that there is no user interaction with an action. A dialog will pop up a screen to force user input, whereas an action is really more of a function call for CRM components to interact with. A dialog forces human interaction, while an action forces workflow or other CRM component interaction.

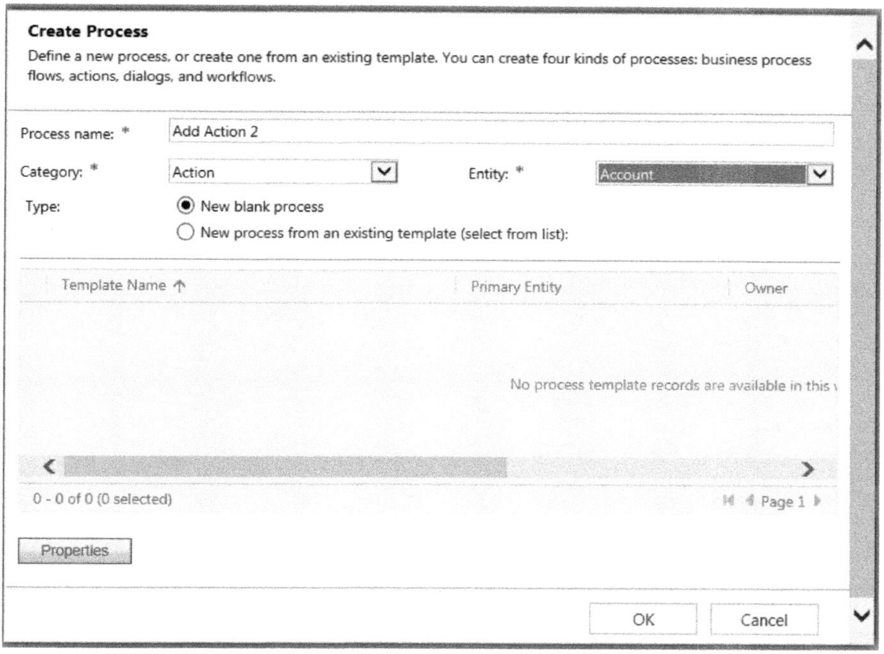

Figure 7.31: Creating an action.

Once you have clicked OK, the configuration screen will open (Figure 7.32). The key items to note and work with are as follows:

- Enabling Rollback
  - If an action fails, this feature will turn on, and undo the changes if any of the action fails. Any transactions performed outside of CRM cannot be rolled back.
  - Determine with the developer whether the rollback must be in transaction or not.
- Workflow Log Retention
  - Keep Logs for workflow jobs that encounter errors
  - Use this to help troubleshoot any issues
- Direction
  - Use this section to define any input or output variables that will be used in the calling process for this action. You can use an action like a method in code to do work and return a value to the calling process.

– The lower section of the window allows you to configure the steps within the action just like you would with a standard workflow.

Actions can call workflows, other actions, and workflow activities. Workflows can call actions as well!

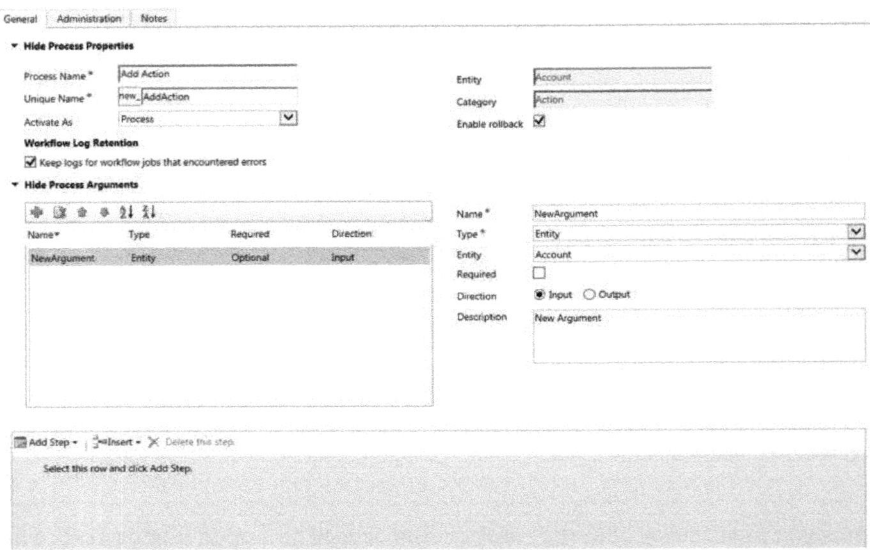

Figure 7.32: Configuring the action.

Summary

At this point you have a good understanding of how to configure a basic workflow in your environment. You should also have a basic idea of what is involved in building out more complex functionality, whether with the standard configuration options within the workflow step configuration, or through the use of workflow activities. You also have familiarity with two other process types: Dialogs and Actions. As you build out your CRM environment, you will find many uses for workflows and will come to depend on them to update fields, send notifications, create or update entities, perform calculations, call external systems, and all manner of other things.

Chapter 8
Solutions

Solutions are a critical feature of Dynamics 365. A solution is a collection of changes to a CRM environment that can be exported and then imported into a new CRM organization. This allows for movement of functionality from Development to Production. It also allows for managed technology change control. This chapter will give an overview of how to create a solution, how to export that solution, and how to import the solution into a new CRM environment.

Some environments will use only the Default solution, while others will dedicate one solution to each developer to keep some visibility into what individuals are working on and deploying. Understand that any change to a component within one solution will cause that change to occur to the same component in another solution; this means that the solutions share the pool of common components.

Creating a Solution

To create a solution, begin by navigating to Settings and then click Solutions. The screen shown in Figure 8.1 will open.

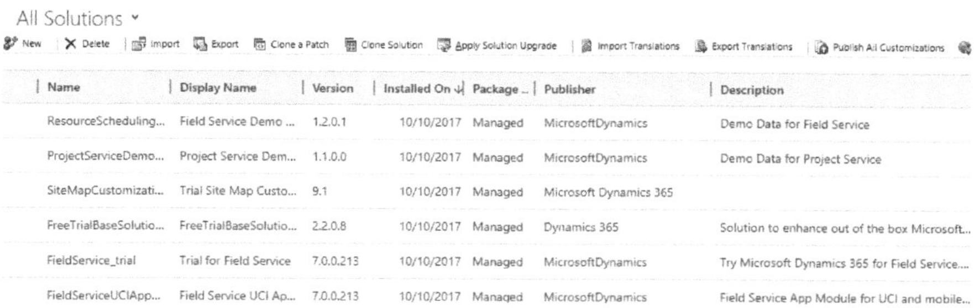

Figure 8.1: Show All Solutions.

Clicking on the New button in the menu ribbon will open the box window shown in Figure 8.2. You will be able to enter the following information.

DOI 10.1515/9781547400065-008

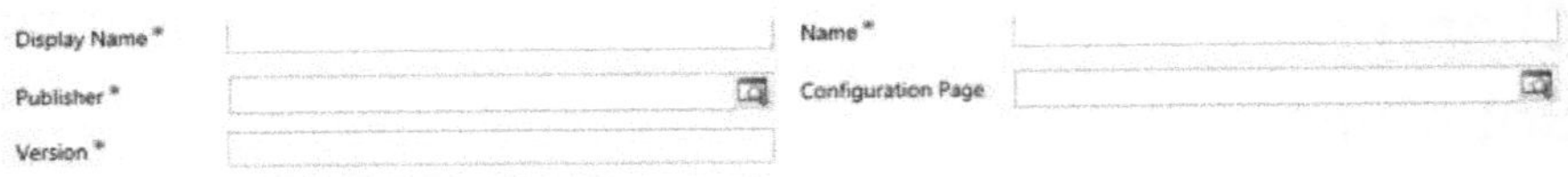

Figure 8.2: Creating a new solution.

1. A Display Name and Name. Note that when tabbing out of the Display field, the Name field will be automatically populated with the same value entered into the Display Name. It can be changed.
2. Publisher. The way to envision a publisher is as enabling a tracking mechanism for who made what changes. When a publisher is assigned, it contains a prefix that will precede any CRM component change. This allows a company to assign a publisher to each developer or CRM Admin to track changes. You will likely want to create a new Publisher when you are working in a new CRM environment. If you don't, the prefix will always default to "new_". To create a new publisher, take these steps:
    a. Click the magnifying glass in the property box to open the configuration screen, then click the Look Up More Records link, then click the New button in the bottom left of the dialog box. The Publisher Information dialog box will display (shown in Figure 8.3).
    b. Enter a display name, name, description.
    c. The most important property is the Prefix. This will precede the name of any new CRM component that you create. This makes it easy to track who created the modification, and should be used when working with vendors or multiple development resources. Each solution can have a different publisher.

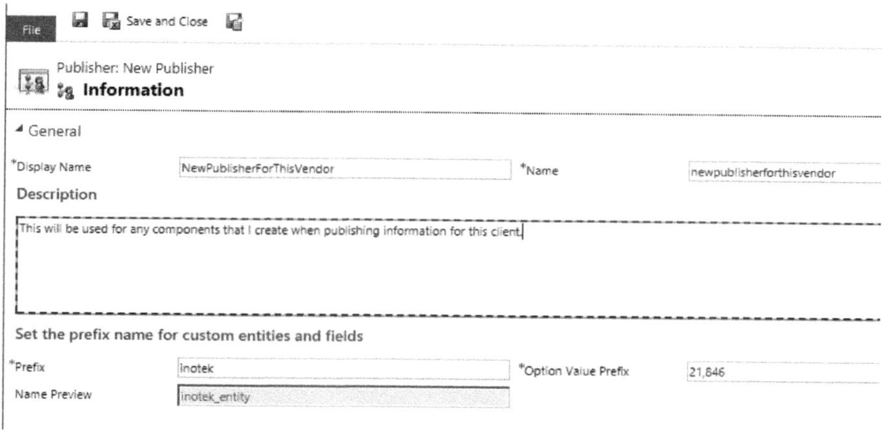

Figure 8.3: Creating a new publisher.

1. Version. Use this field to track related information, such as how this solution maps to your project tasks, or how this version compares with another solution that was previously imported into a target environment.
2. Configuration Page. You can set configurable fields in the configuration page for a solution. This allows you to more easily change URLs or other pointers that may be different in target environments. For example, you may have a Development environment where you create the solution, and QA environment where you plan to import it. These may point to a different SQL Server Reporting Services report server, and therefore you may wish to have a configurable field on your solution.
3. Save and Close the Publisher dialog box. Then click Add on the Lookup Record dialog box, and you will return to the New Solution page.

On the New Solution page, after you click the Save button, the solution structure is saved and can now be used. An empty solution will be displayed, like the one shown in Figure 8.4. This screen functions in a very similar way to Default Solution changes. Any New or Existing components will appear under each of the respective Components in the left-hand navigation menu.

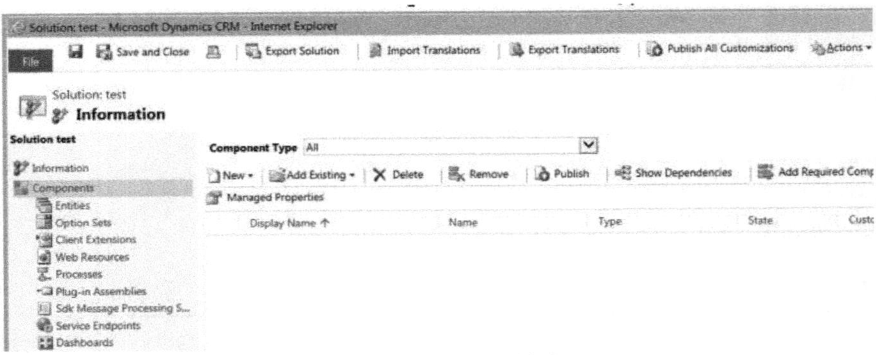

Figure 8.4: Currently this solution has no components added to it.

For completely new CRM components, click the New button. To make changes to existing CRM components, click the Add Existing button and select from the list of existing components, as shown in Figure 8.5. When you select the component type, a list of those components will be displayed.

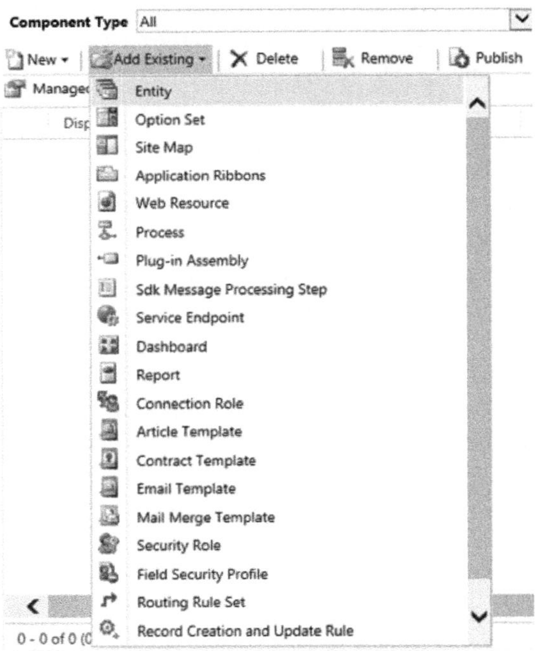

Figure 8.5: Adding components.

After you have added the components that you want to your solution, you will need to publish those components. The act of publishing causes all changes in the new or modified components to be made available to users in the environment. To publish individual changes, click on each component that you want to deploy and click the Publish button. Figure 8.6 shows that account entity has been selected and the Publish button is clicked on the toolbar above it.

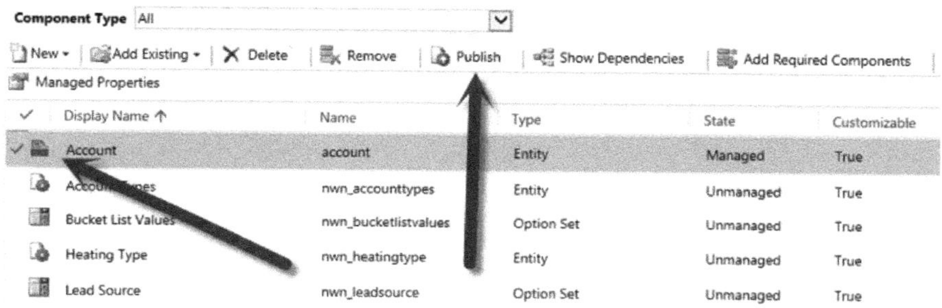

Figure 8.6: Publishing changes for a single entity.

If you select more than one component to be published and click the Publish button, you will get the error shown in Figure 8.7. You can only publish one component at a time when publishing individual components. Use the Publish All Customizations on the main toolbar if you have many changes that need to be deployed.

Take care with the Publish All Customizations option. This can cause issues if there are components that have had changes that are not ready to be made public.

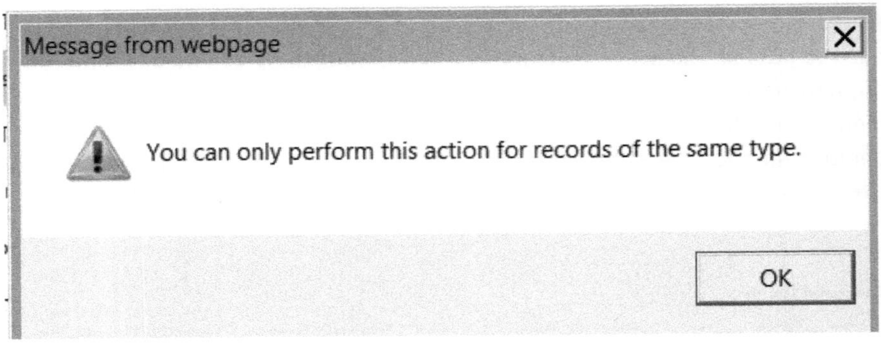

Figure 8.7: Error when trying to publish changes for more than one entity at a time.

You can continue to add components within your solution. Once complete, you can move on to exporting your solution.

Exporting a Solution

Once all changes in a solution have been published, you can export the solution for importing into another CRM environment. Typically, development is first done in a development environment, then exported and imported into a test/QA environment. After testing has been completed, the solution is imported into the final production environment. To export a solution, take the following steps:

1. From the main menu of the solution, click on the Export Solution button. The warning screen shown in Figure 8.8 opens. You are not required to publish all customizations prior the export, but only published changes will be included in the exported solution.

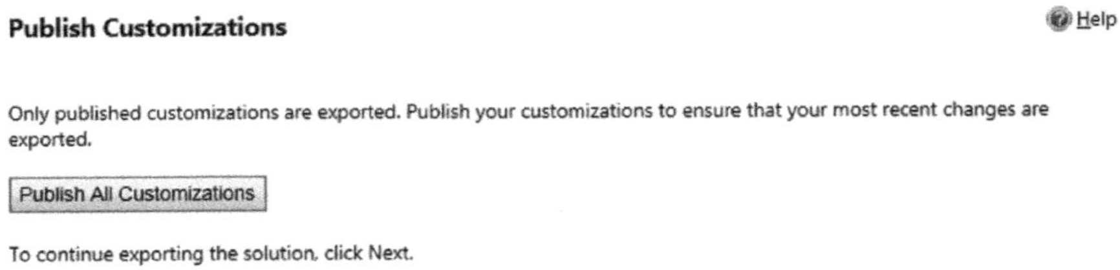

Figure 8.8: Option to publish all customizations prior to export.

2. Click the Next button and CRM will move on to the next screen where it will run a comparison of the changes in the solution. The check will display a list of missing components that if not already in a targeted environment will cause the importing of the solution to fail. You do not need to include these in your exported solution, but when you go to import them into another CRM environment, they must already be present in the target environment. Figure 8.9 shows this screen. You can click Cancel to go back and add them to your solution, or you can click Next knowing that they already exist in the targeted CRM environment.

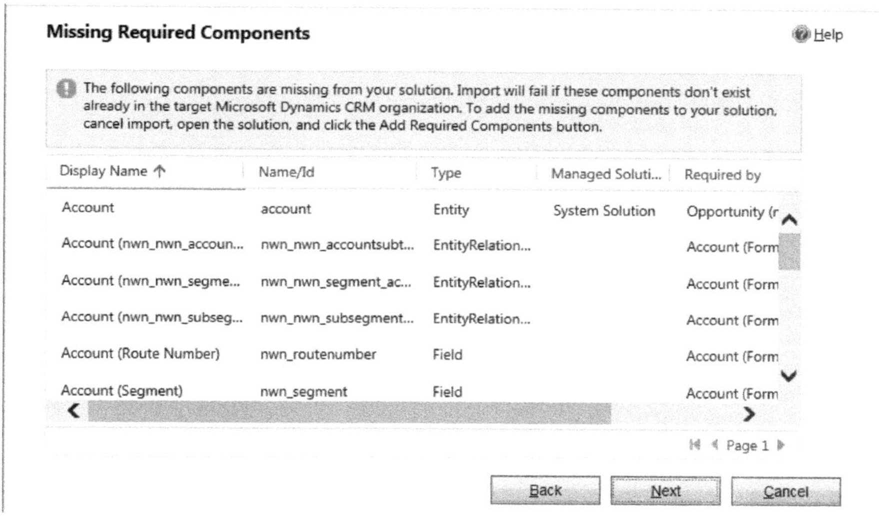

Figure 8.9: A warning that some components are referencing other components not included in the solution.

3. Once you click Next, the screen shown in Figure 8.10 will open. Add any feature in the list by checking the box next to it. Frequently, these boxes are left unchecked, as many CRM environments simply use the default settings for these components. If, for example, you have configured auto-numbering for one of your entities, you'll want to check the Auto-numbering box.

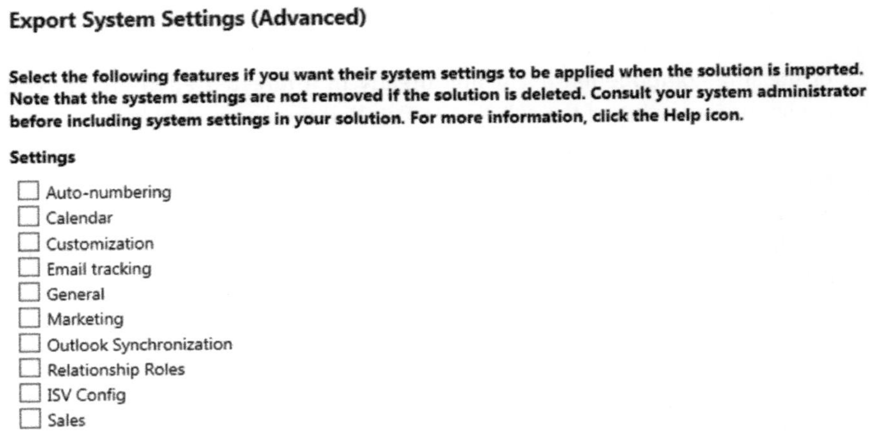

Figure 8.10: If you want to, you can include system settings in your export.

4. Clicking Next on the previous screen will open the Package Type dialog box, and ask you to choose whether you will be exporting a Managed or Unmanaged solution (see Figure 8.11). Some key items to note on these two types are:
   a. Unmanaged
      i. All Solutions must begin as Unmanaged. This enables changes to be made to the solution after it is imported into another environment.
      ii. Deleting the Unmanaged Solution only removes the solution. It does not remove the component changes made by the solution.
      iii. When you are doing development of a CRM solution that will remain within a single organization, you will typically export an unmanaged solution. You are not trying to protect your code, and you want to be able to modify these solutions as you move them between environments internal to your organization.
   b. Managed
      i. Since all solutions start as Unmanaged, the only way to create a Managed Solution is to create it during the export process. This is to ensure no component changes can be updated. It also allows for the uninstall of the changes made

by the solution. This keeps the solution feature working and prevents modification to the functionality.

ii. Removing a managed solution will not only remove the solution, but also all changes made in the solution.

iii. When you are developing solutions that you will distribute to third-party environments, it is common to deliver managed solutions. For example, if you are selling a plugin to a number of organizations, and don't want them to make changes to your configuration and settings, you will export your solution as managed.

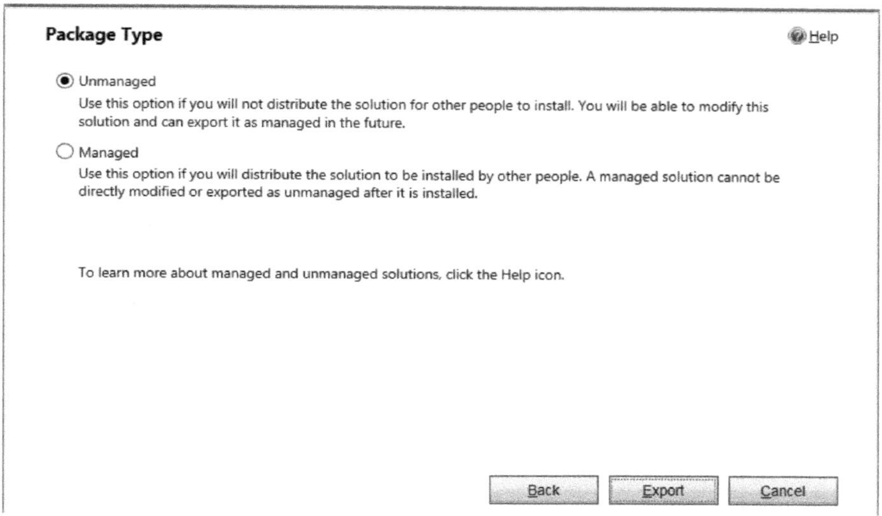

Figure 8.11: Determine whether you are exporting a managed or unmanaged solution.

5. The Package Type screen is the final screen, and once you select Managed or Unmanaged, you can click the Export button. The export process will create a zip file with a number of files in it and can take a while to generate. Larger CRM solutions can take 5-10 minutes to create the export, while smaller ones will be almost instantaneous. When the export is complete, a ribbon at the bottom of your browser window, similar to that shown in Figure 8.12, will be displayed.
   a. Click Open to open the zip file. This has limited value, but will show you the many files that are included in your export. Occasionally

you may want to open these individual files to troubleshoot import errors or to scan through for specific customizations like JavaScript.

b. Save or Save As. By clicking Save, the default directory is Downloads. To change the directory, click the arrow next to Save and choose Save As.

c. Cancel. This will stop the export.

For this exercise, click Save, which will save the file to your Downloads folder so you can use it as the import file in the next step.

Figure 8.12: Don't miss the ribbon at the bottom of your browser.

Importing a Solution

Importing a solution will allow you to take an exported file and import it to a new environment. This allows you to easily move changes and updates between environments without having to manually make changes. To import a solution, you can take the following steps.

1. From the toolbar on your Solutions window (shown in Figure 8.13), click the Import button.

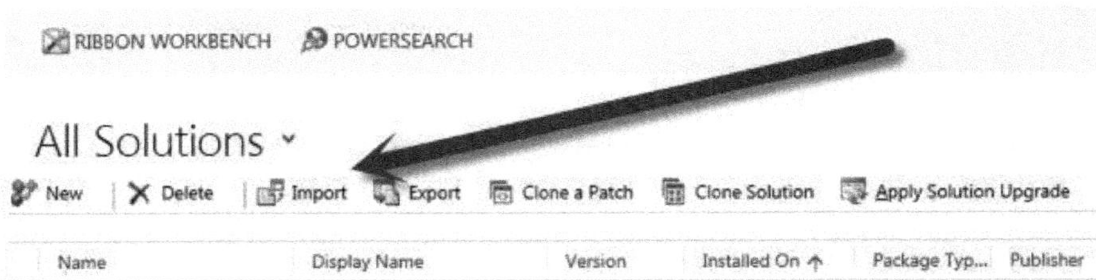

Figure 8.13: The Import button.

2. Once you click the Import button, a screen will open (Figure 8.14) prompting you to browse to the location of your import file. This will be the full .zip file that you previously exported. Generally, you will import a solution into a different environment, but there may be times when you want to export a solution from one environment and import it right back into that environment.

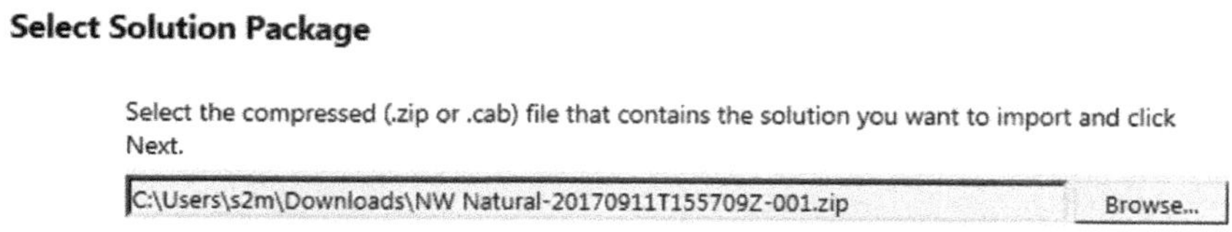

Figure 8.14: Browsing to the .zip file.

3. Clicking the Next button on the Select Solution Package screen will take you to a confirmation page where you can view the package details, if desired (see Figure 8.15).

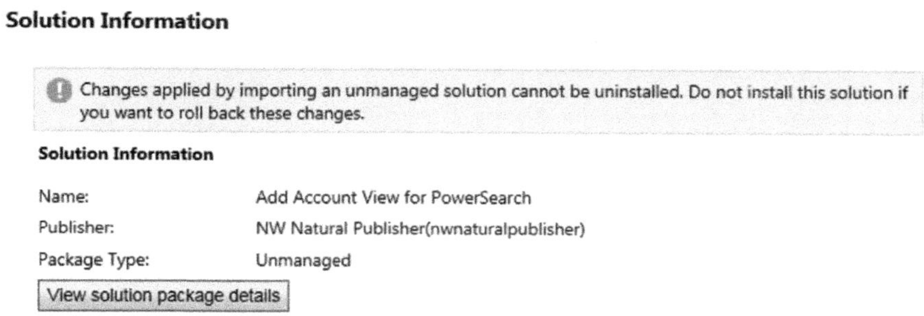

Figure 8.15: The overview of what you are about to import.

4. You can now click the Import button. A progress indicator will appear (Figure 8.16).

The movement of the progress bar at this stage is not always a good indicator that you are nearing the completion of an import. However, if the progress bar does not move for a length of time, you are very likely going to get an error that you will have to troubleshoot. Typically, errors are due to referenced components in your imported solution that neither exist in the solution nor in the targeted environment.

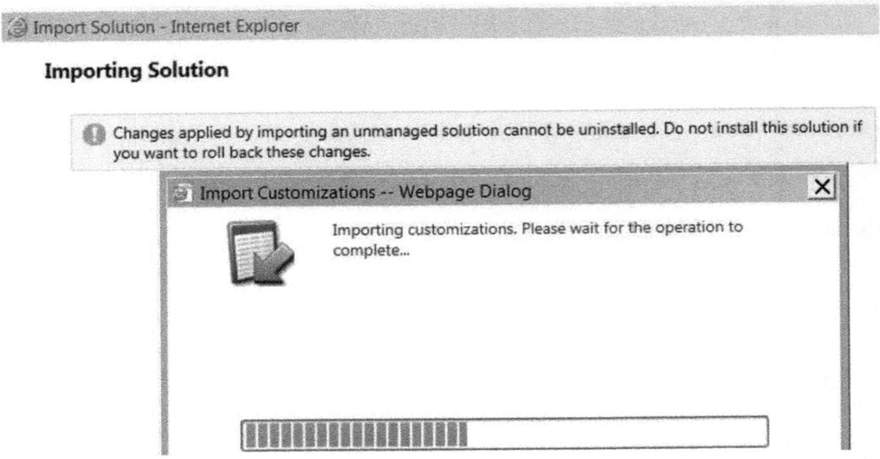

Figure 8.16: The progress indicator doesn't provide much help in determining how far along you are with the import.

5. When the import has completed (either successfully or with failures), you will get a summary screen (Figure 8.17). The screen will give information about each of the components imported, along with success, warnings, and potential complete failure of the import. If you do get a failure, click the Download Log File button.

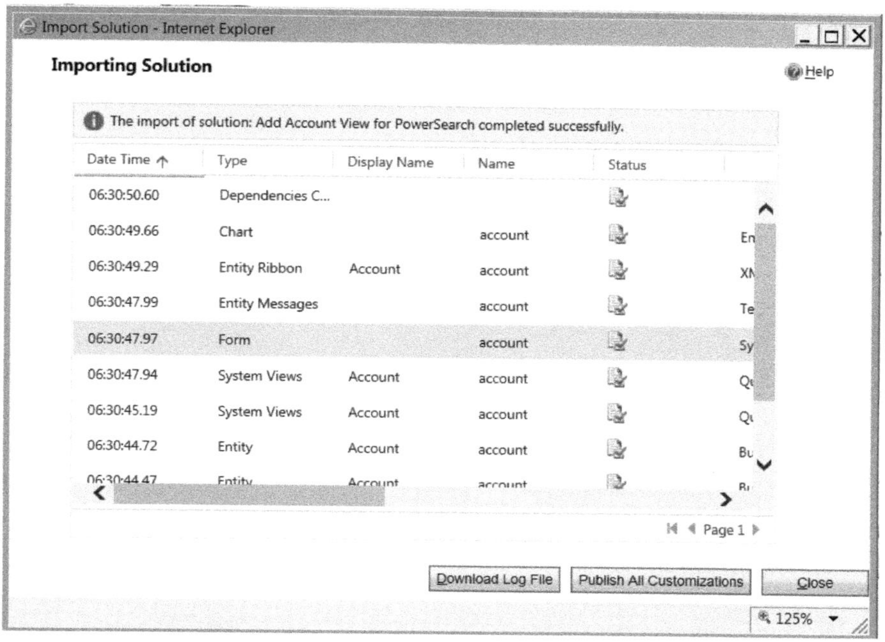

Figure 8.17: The summary page indicating success or failure of your import.

6. If everything goes well with your import, you will need to complete the process by clicking the Publish All Customizations button. This will ensure that everything that just imported will be made public.

Solutions and Versioning

Solutions also provide a way to back up work. Many organizations use TFS (Team Foundation Server) or other code repositories to track changes and keep backups. Exporting a solution with all your components in it and checking it into TFS will ensure you have the ability to set up new environments at will.

If you decide to use solutions for version control and backups, there are several things you want to do:

1. Ensure your solutions are named in an intelligent way. For example, add a version number at the end of the file name along with a date. It could be something like ReportSolution20170511-1.0.0.5.
2. Add details in your description tying it to a Sprint cycle that you may have had (with Agile methodology) or based on the specific developer.

Figure 8.18 gives an example of how the information can be entered on the solution to give details about what occurred with this solution. This type of approach is critical in a multi-developer environment to keep track of who is doing what and when.

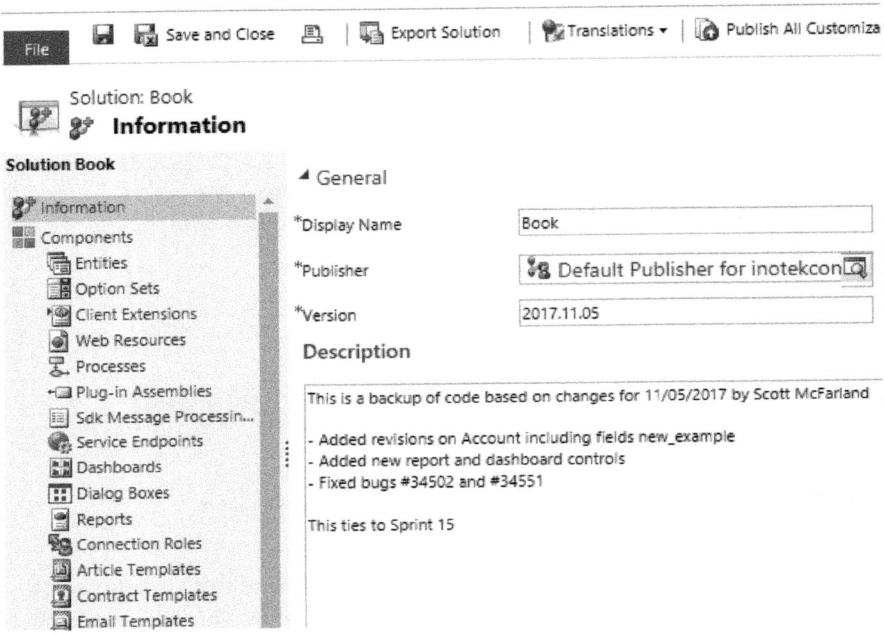

Figure 8.18: Use the information tab of your solution if you are planning on versioning your solutions.

Summary

Solutions serve primarily as ways to move your components from one environment to another. If you have a single environment, then you'll just use the default solution. If you have multiple environments, you may decide to separate your solutions by developer or type (for example, you may have one solution for reports, one for entities, and so forth). This chapter should have supplied you with the basic knowledge you need to configure, export, and import your solutions.

Chapter 9
Templates

Templates allow you to pull data from a record with a basic outline of information already in place. This can be very useful when you have many artifacts that must be created that all follow a similar pattern. This chapter will show how to create and use an email template. Dynamics 365 offers templates for the entity types shown in Figure 9.1.

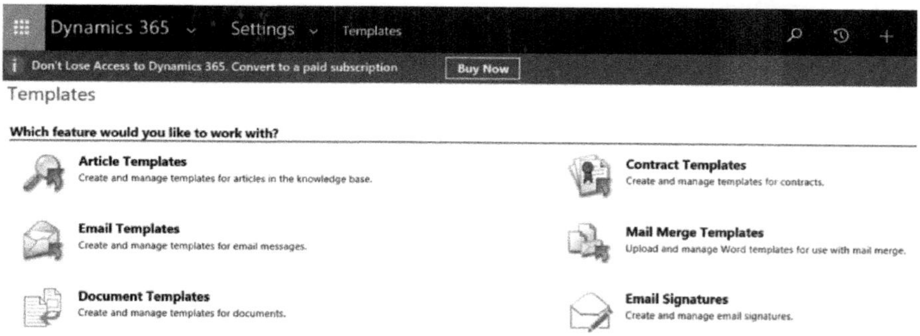

Figure 9.1: The various types of templates available in D365.

Creating a Global Email Template

To create an email template, begin by navigating to Templates on the main Settings menu. Once there, click on the Email Templates link. A list of existing templates will be displayed, along with a toolbar where you can click the New button, as shown in Figure 9.2.

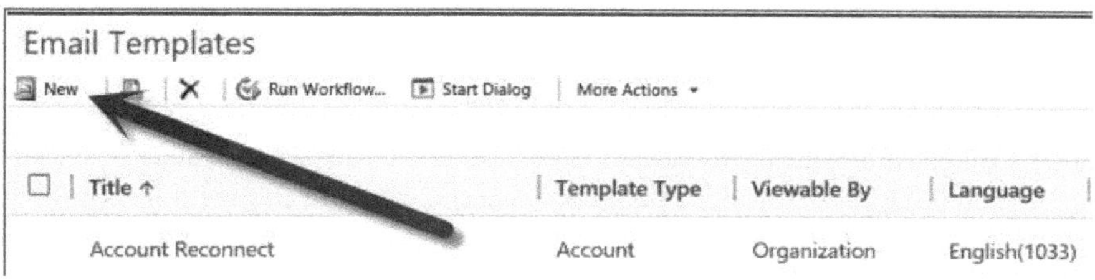

Figure 9.2: Click the New button from the list of existing email templates.

DOI 10.1515/9781547400065-009

When you click the New button, you will be prompted to select the Template Type (see Figure 9.3). You can choose either Global or a specific entity. By choosing Global, the template will be available from any entity. If you choose a specific record type, the template will only be available from within that entity.

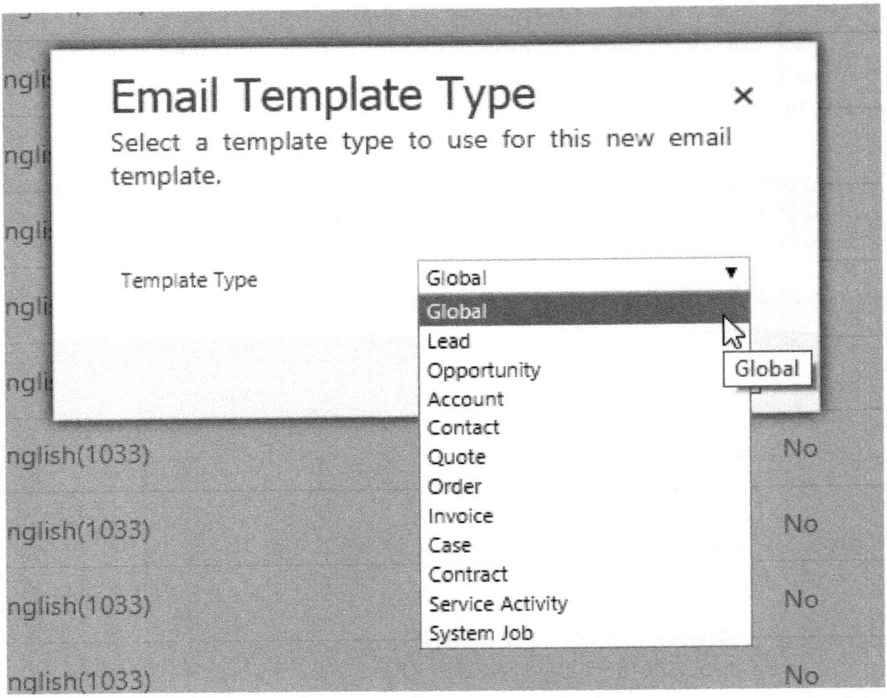

Figure 9.3: Determine from where your template will be accessible.

After you choose the Global template type and click OK, you will enter values in the Title and Description fields. In the Subject and Email body sections, you can type static text, or bring in dynamic values from the User entity. For example, as shown in Figure 9.4, you can type in a partial static subject line followed by a dynamic User value. Clicking the Insert/Update button on the ribbon toolbar will open a dialog box where you can add specific fields—but only from the User entity. In the Data Field Values dialog box, click the Add button to add the applicable dynamic values. Organize the values as needed, and then click OK.

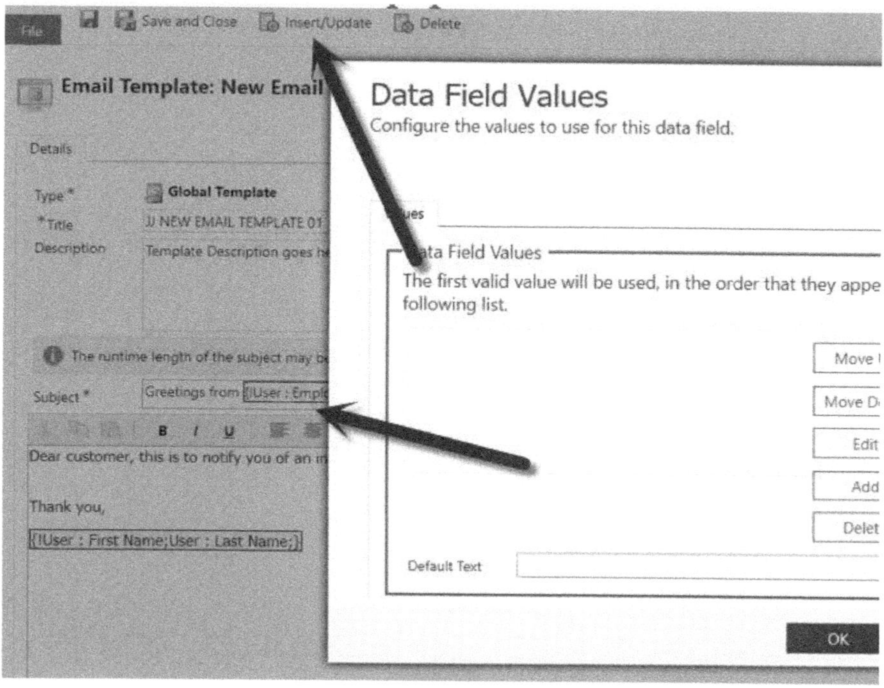

Figure 9.4: You can add dynamic values to the subject and email body by clicking Insert/Update.

Figure 9.5 shows a completed Global email template's description line. As you can see, once again you are limited to dynamic values from the User entity. This is because CRM will not have access to a specific entity to pull data from. If you want to be able to access fields from an individual entity, you must create an entity-specific email template.

Figure 9.5: Global email templates can only access User information for dynamic values.

Creating an Entity-Specific Email Template

If you desire to add more fields to the email template, a specific-entity template must be selected from the Email Template Type menu (instead of the Global option shown above). As an example, you can create an email template specific to the Account entity. Following the same steps as already described for the Global template, select the Account entity as the template type (Figure 9.6).

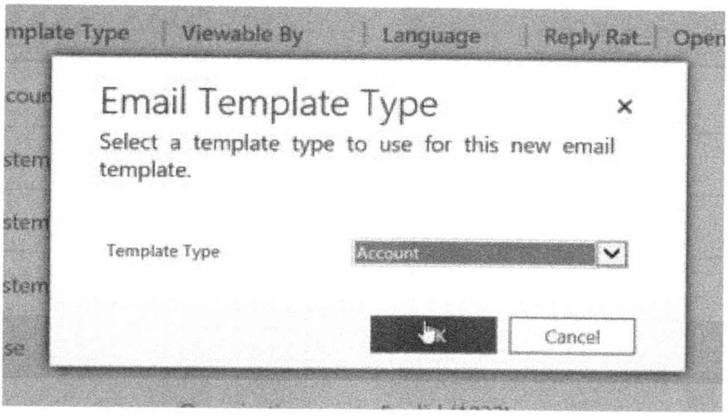

Figure 9.6: Selecting Account as the specific entity email template.

The same configuration screen will appear, but this time you will have access to the dynamic values available for Account and related Contact information. When you click the Insert/Update button, and click the Add button on the dialog box that opens, you will be given the option of which entity and related fields to choose from (Figure 9.7).

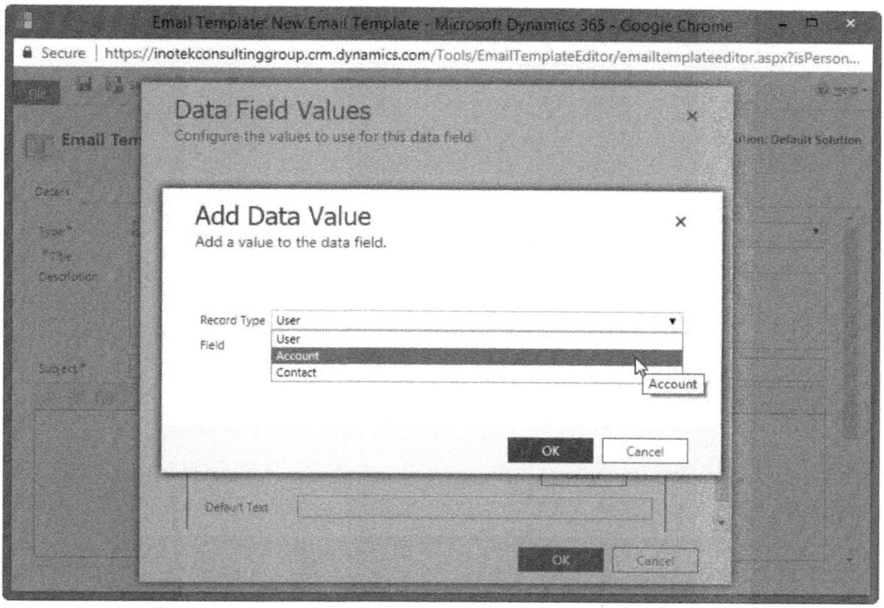

Figure 9.7: You can select from several entities and fields.

Figure 9.8 shows an example of an email template that has been configured with information from both the account's contact and related user information. When you have finished composing the subject and body of your email template, you can click the Save and Close button. You'll now be able to access your template from within CRM.

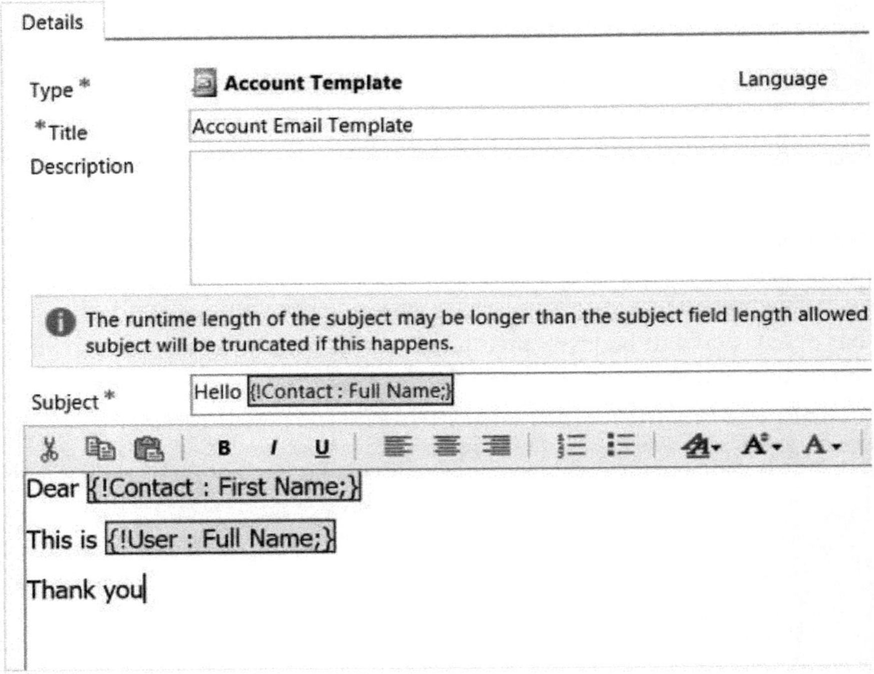

Figure 9.8: Pulling data in from several entities on the template.

Using an Email Template

Once you have saved your template, you will be able to access it from within Dynamics. For the account-specific email template, open an existing account record and click on Activities and then Email (shown in Figure 9.9).

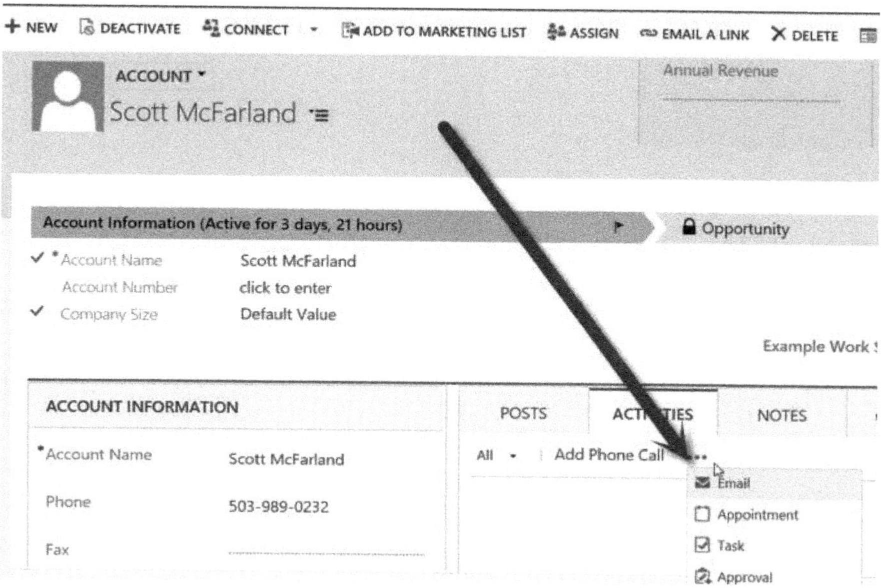

Figure 9.9: Click Email from within the Activities sub-menu.

On the email window that opens, click the Insert Template button. This will open a screen where you can select the template you wish to use. Notice that in the example shown in Figure 9.10, there are two templates that have been created, one appearing under the Global Templates and one under the Account Templates.

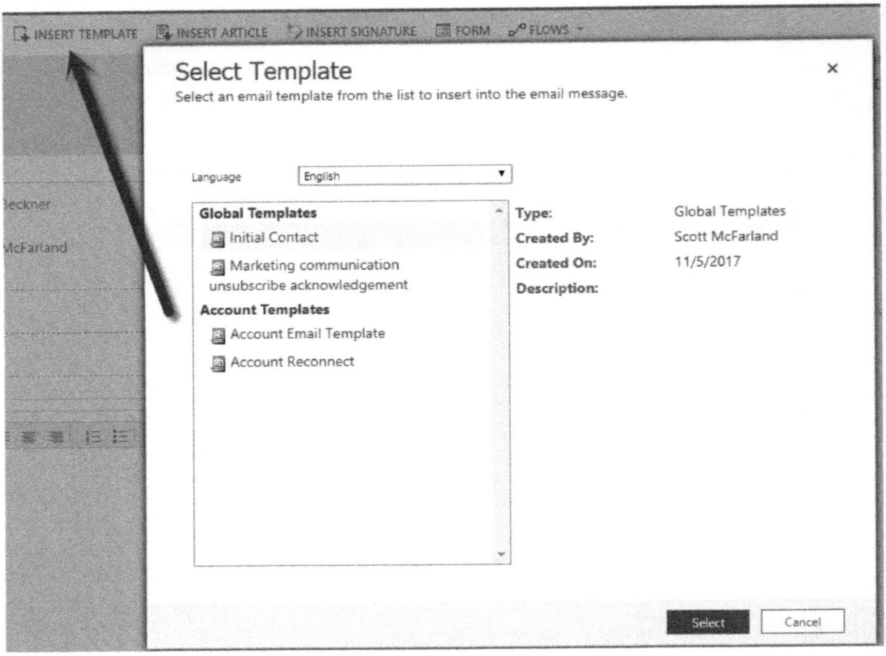

Figure 9.10: Click the Insert Template button and select the template you want to use.

In this example, when the Account Email Template is selected, it will override the blank email with the text and fields from the template (shown in Figure 9.11). You can make further alterations to the email before sending it. When you are ready to send, click the Send button on the toolbar and your process is complete.

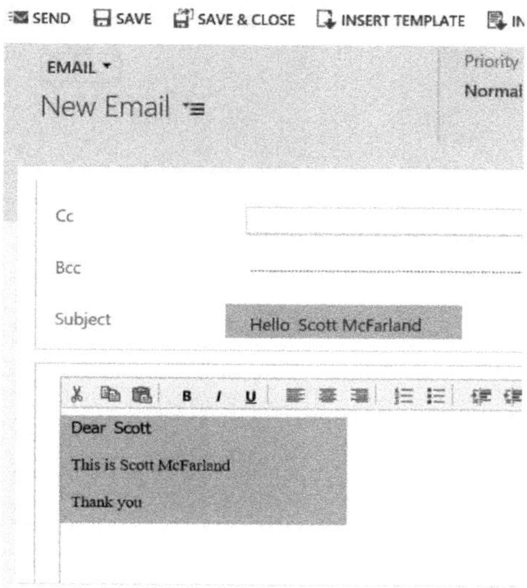

Figure 9.11: You can alter the email before sending it.

Other templates can be created in a variety of locations. The easiest place to create new templates for articles, contracts, and mail merges is directly out of the solution. You can do this by click on Components, and then selecting the template you want to create. Figure 9.12 shows the existing Article templates, and where a new Article template could be created from.

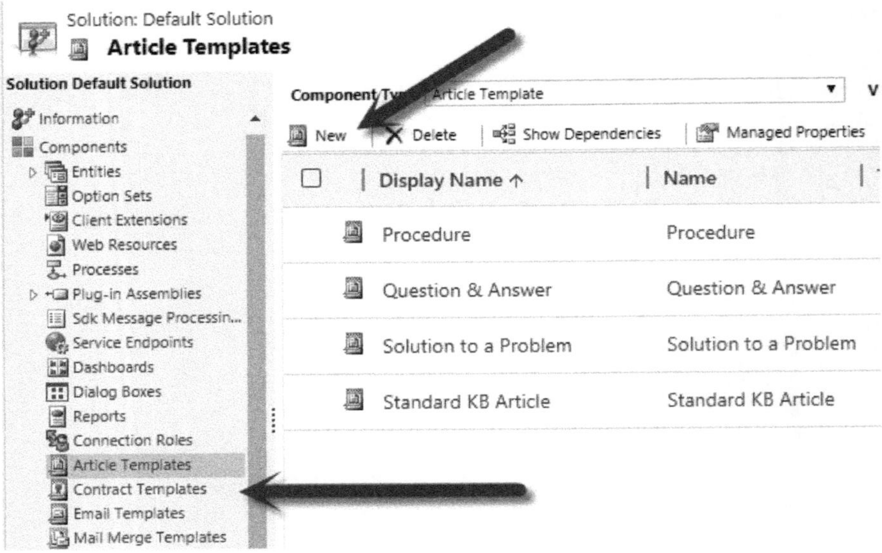

Figure 9.12: Adding a new Article Template.

An Article template allows you to force a standard structure to all articles that are available within Dynamics. Organizations that open their Dynamics environment up to customers may want to publish weekly articles on how to sign up for certain events (for example). Each week, a User of CRM could start with an Article template and then publish it. By utilizing a template, each article will have the same look and feel.

Summary

Creating templates is very straightforward. The ability to create a starting point for various entity types through the use of templates can save users time when doing repetitive tasks. While there are several entities that templates can be created for, the most common is the Email template. Creating an Email template is fast and easy to do, and will ensure that users are sending email that match a common layout and standard information when sending to external resources.

Chapter 10
Dashboards

Dashboards are ways to display collections of information and data that allow a user to quickly assess information across entities. A dashboard is a set of components that can pull data and display information in a variety of formats. Dynamics 365 comes preconfigured with a generous number of dashboards that you can utilize. This chapter will outline how you can modify an existing dashboard, create a new dashboard, and build custom components like charts and views to add to your dashboard.

Setting Your Landing Page

When you first log into Dynamics, you will land on the Dashboards page. You can change your landing page by clicking on the gear in the upper right of your browser and selecting Options (see Figure 10.1).

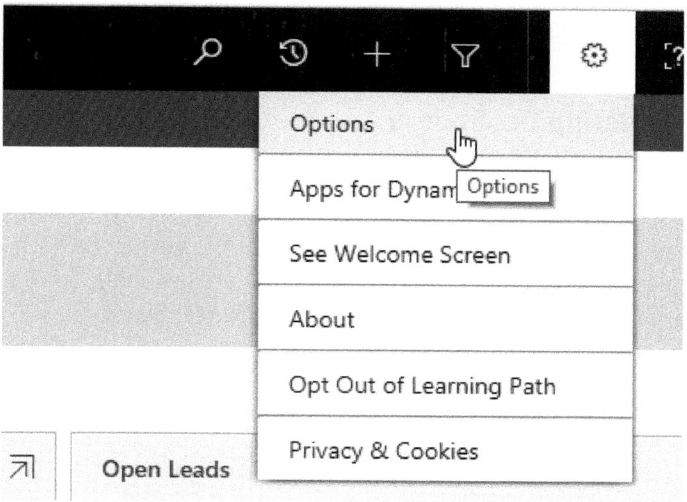

Figure 10.1: Accessing personal options from the main screen's toolbar.

On the Options page, you can set the Default Pane and Default Tab properties in the General Settings, which will change where you initially land. Note that this

DOI 10.1515/9781547400065-010

approach will change it for the current user logged in, and will not impact other users. In Figure 10.2, you can see that these properties are being changed so that the default location the user will land in will be "Opportunities."

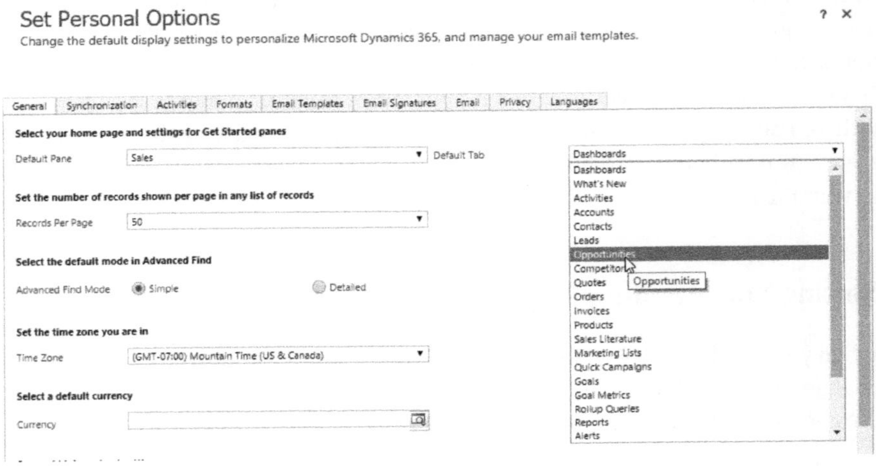

Figure 10.2: Altering your default landing page.

Customizing an Existing Dashboard

You can customize an existing dashboard easily. As you can see in Figure 10.3, there are many dashboards available to select from that ship with D365. The default is the Sales Activity Dashboard, clicking drop-down arrow next to the dashboard title will display a list of the other dashboards that are available. In this section, we'll make some modifications to this Sales Activity dashboard.

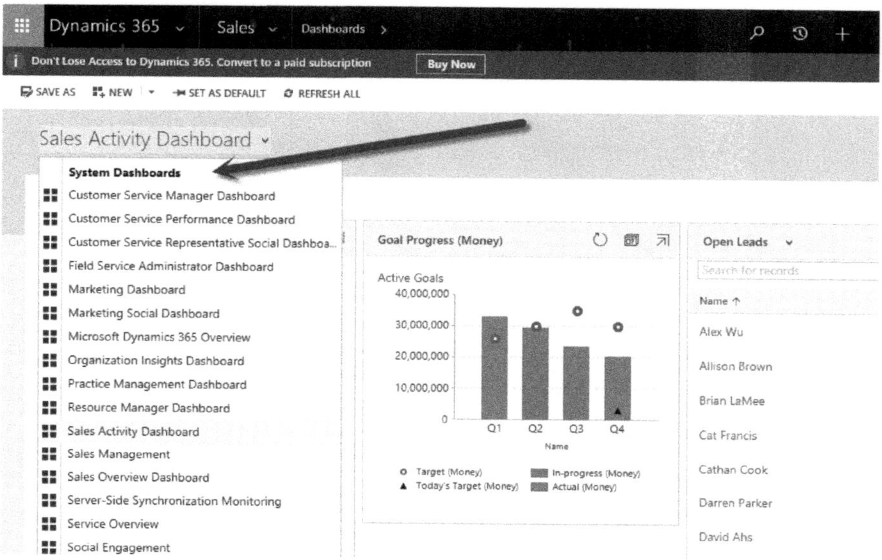

Figure 10.3: The numerous dashboards available in Dynamics 365.

To modify an existing dashboard, open your solution and click on Dashboards (see Figure 10.4). Scroll down to the dashboard you are interested in modifying (in this case, the Sales Activity Dashboard) and double-click it. This will open a new window where you can make modifications to the layout and content of this dashboard.

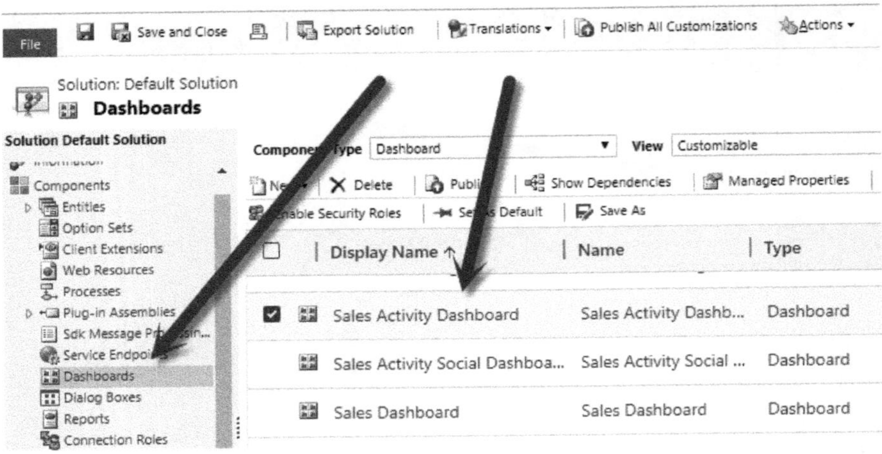

Figure 10.4: Modify a dashboard from the solution.

In the screen that opens, you will see a new toolbar with a number of options on it. You will also see that you can change the name of your dashboard, and can alter all the components that display in the dashboard (Figure 10.5).

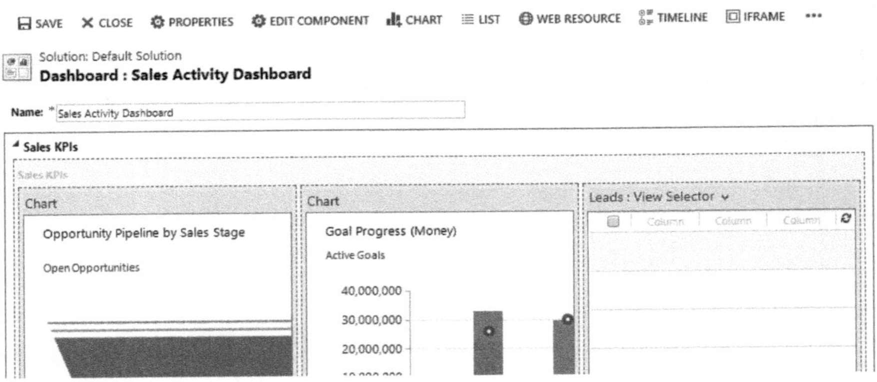

Figure 10.5: The dashboard editor screen.

To modify the current dashboard, take these steps:

1. Remove the Leads: View Selector by clicking that pane in the main dashboard area and then click the REMOVE button on the toolbar.

2. Click on the main Sales KPI section and click the PROPERTIES button. Change the label of this to read "Sales Information" and select the "Show the label of this section on the Dashboard" property. Click OK when you have altered this information (Figure 10.6). You have to click on the section within the editor in order for these properties to show.

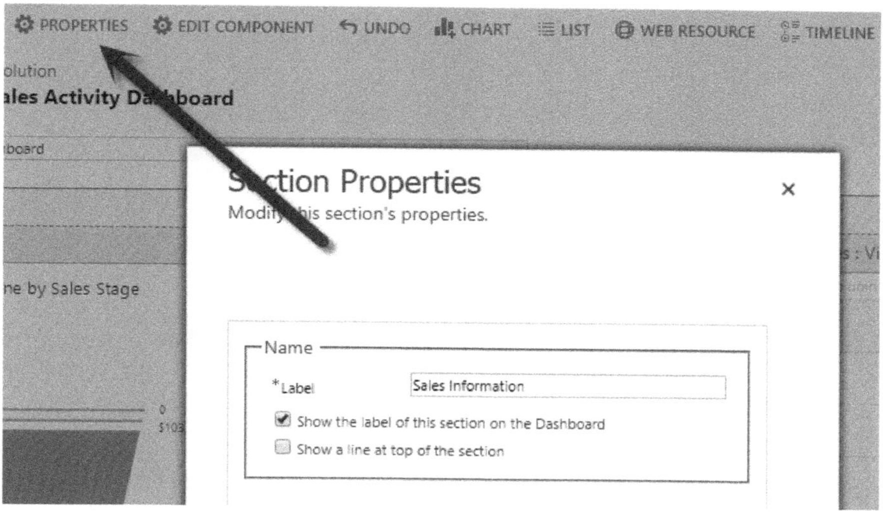

Figure 10.6: Showing the properties on the section that holds the various charts.

3. Click on the chart tile "Goal Progress (Money)" and then click the EDIT COMPONENT button. In the Chart Options section of the screen that opens, change the Default Chart to Percentage Achieved and click OK.
4. Add a new chart by clicking the CHART button. In the window that opens (see Figure 10.7), select "Lead" as the Record Type, "All Leads" as the View, and "Incoming Lead Analysis by Month" as the Chart, and then click the Add button.

You can change where a control is located by dragging and dropping it. However, you may find that the browser you are in doesn't allow you to move these tiles. If you have trouble dragging and dropping tiles, try a different browser, or try closing out of the browser session you are in and reopening.

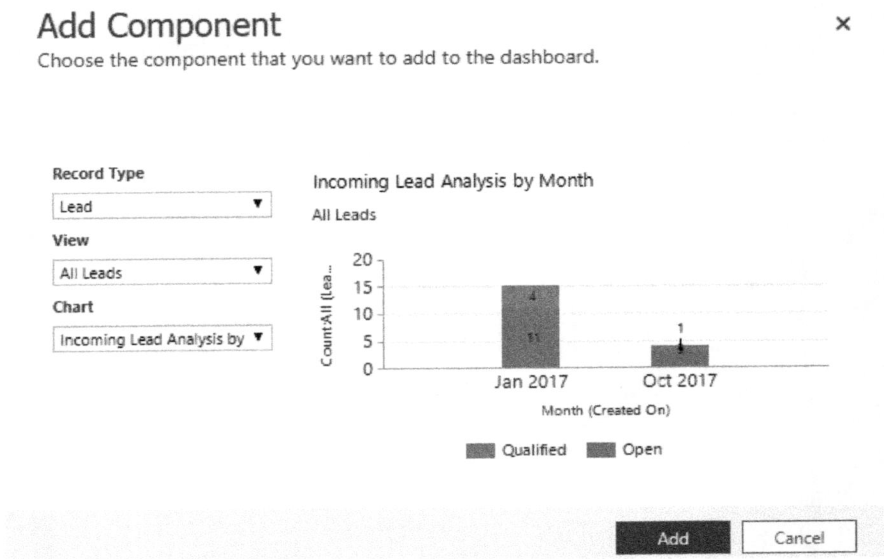

Figure 10.7: Adding an existing component.

When you are done altering this existing dashboard, click the Save and then the Close buttons and return to the main dashboard list. Click the checkbox next to the dashboard you just changed and click the Publish button on the toolbar (see Figure 10.8).

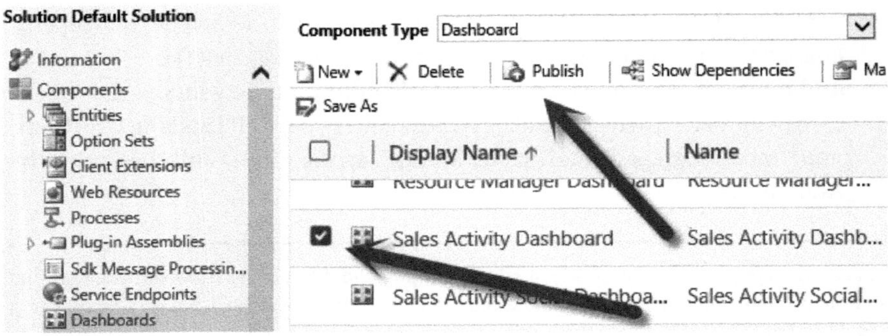

Figure 10.8: Publishing the changes to your modified dashboard.

Creating a New Chart

To create a new chart to add to a dashboard, you must add it to a specific entity. To demonstrate, we'll create a new chart on the User entity. Expand the User entity in your solution, and then click on the Charts tab. In the view on the right, click the New button (see Figure 10.9). This will open the chart editor in a new window.

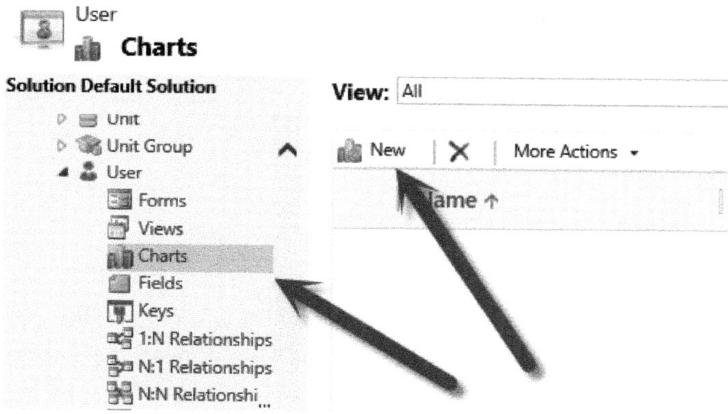

Figure 10.9: Creating a new chart.

The chart editor window that opens is a very detailed editor that allows for the creation of a variety of charts. Every chart that you create will be based on an underlying view. If you want to create a chart that is based on data that isn't already in a view, you will need to create a new custom view before proceeding (see the end of Chapter 3 for instructions on creating and customizing views).
Start by giving your chart a name. We'll build a chart that shows users created over time, call this the User Creation Timeline, and base it on the Enabled Users view, as shown in Figure 10.10.

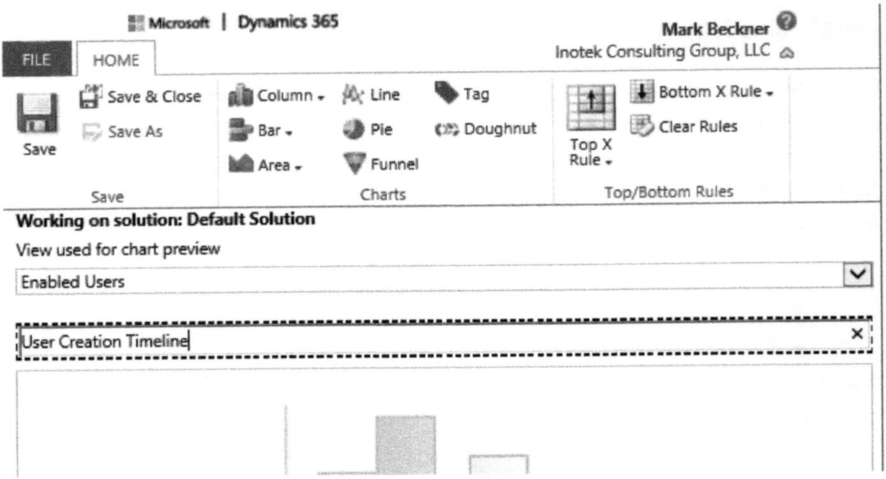

Figure 10.10: The chart editor.

Now click on the Bar icon in the editor toolbar and select the Bar chart. This will allow you to configure the remaining properties on your screen. A simple configuration of this is to set the Legend Entries to Full Name and Count:All, and the Horizontal property to Created On and Month. The end result will show users that have been created in each month. Figure 10.11 shows this configuration.

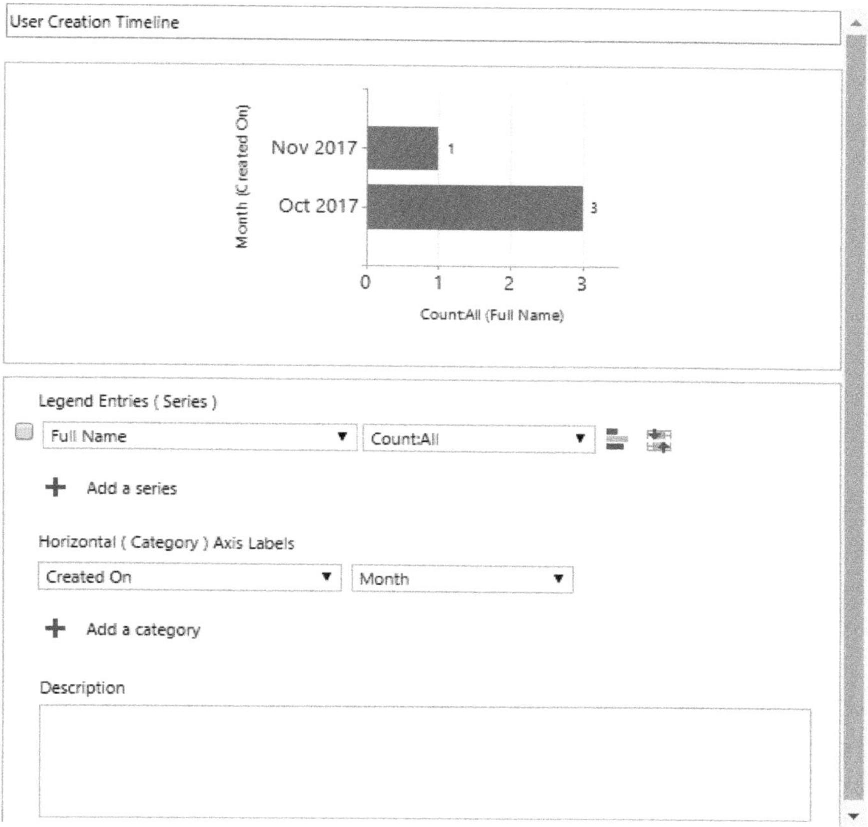

Figure 10.11: Configuring the bar chart properties.

When you have completed your chart, click the Save and Close button at the top of the window. You will now need to add this chart either to an existing dashboard or a new dashboard.

Creating a New Dashboard

We'll add the chart you just created to a new dashboard and, in the process of this creation, demonstrate other functionality available when creating a dashboard. To create a dashboard, take the following steps:

1. Click on the New button from the Dashboard tab in your solution. You will be able to choose the layout of your dashboard from the screen that opens, as

shown in Figure 10.12. Select the desired layout, and then click the Create button in the bottom-right corner of the screen.

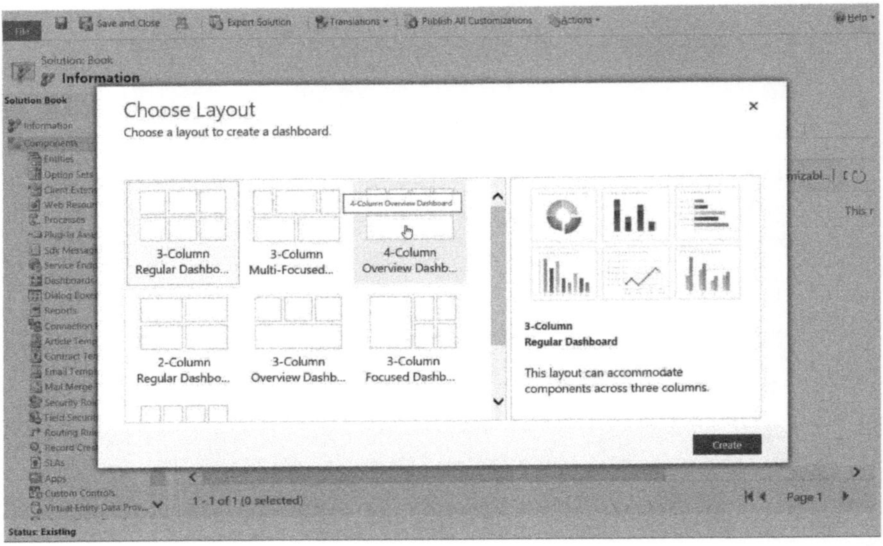

Figure 10.12: Choosing a layout for your new dashboard.

2. When you click Create the main dashboard editing screen will open. Before, when you were modifying an existing dashboard, this screen was already populated. Now it is a blank canvas (see Figure 10.13). Enter a name—for this example, we'll call it User Dashboard.

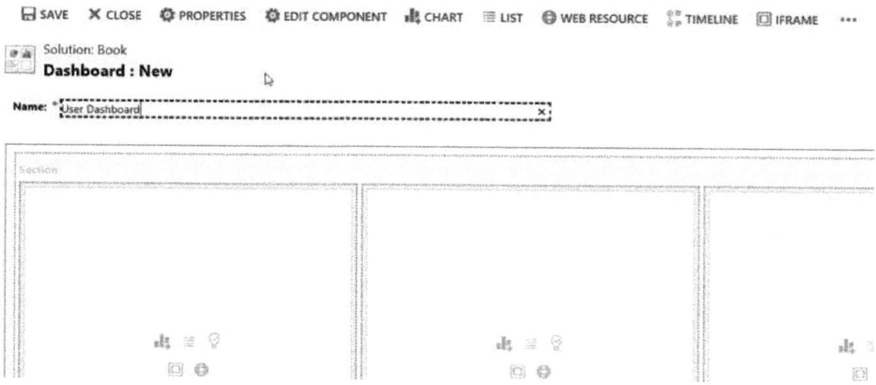

Figure 10.13: The canvas is blank when you create a new dashboard.

3. Add a component. You can choose a chart, list, relationship, IFrame, or Web Resource. For all of these but IFrame, you must have already created the component in order to add it to your dashboard. Add your newly created chart by clicking on the CHART button in the toolbar. Select your chart by choosing "User" as the Record Type in the Add Component window shown in Figure 10.14. Click the Add button when ready.

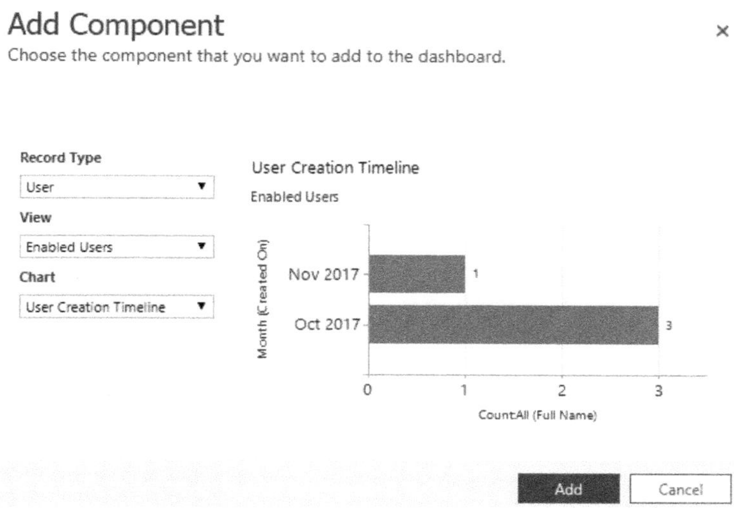

Figure 10.14: Adding the new chart that you created to your dashboard.

4. Back on the main screen, your chart will have been added as shown in Figure 10.15. You can now click the Save button on the toolbar, followed by Close.

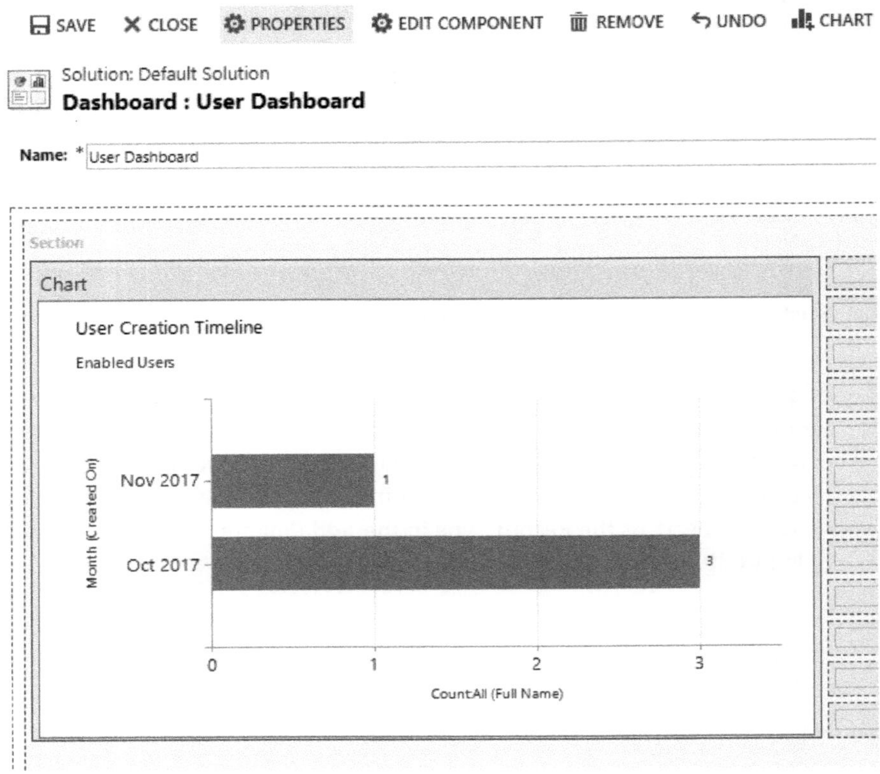

Figure 10.15: The chart added and placed on the dashboard.

5. Back in your main solution screen, you can select your User Dashboard and click the Set As Default button to make this the default dashboard if you would like to (Figure 10.16).

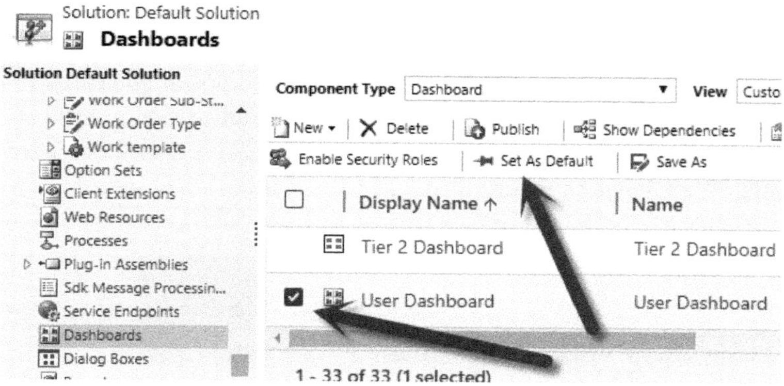

Figure 10.16: Setting your new dashboard as the default dashboard.

6. Your dashboard can be assigned to be accessible by specific security roles to restrict access to it based on user, team, business unit, and so on. To enable security, select the User Dashboard in the list of dashboards in your solution and then click Enable Security Roles. The screen shown in Figure 10.17 will open; from here you can grant permissions as needed.

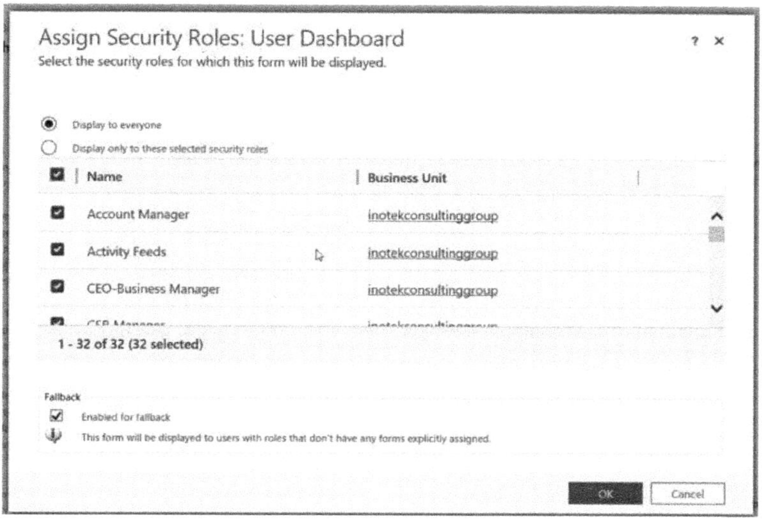

Figure 10.17: You can assign security roles to restrict access to your dashboard if needed.

7. Publish your customizations. You will need to publish the dashboard as well as the User entity in order to have the new chart and dashboard both accessible to users. You can click the Publish All Customizations, or you can publish the individual Dashboard and User entity separately. Once published, and after the browser has been refreshed, you will see your User Dashboard as the default, as shown in Figure 10.18.

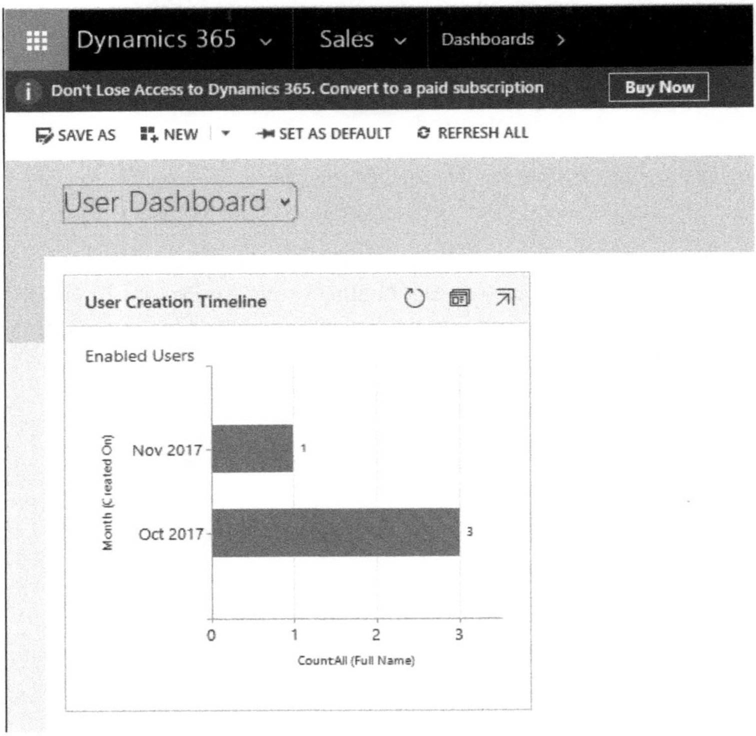

Figure 10.18: After publishing your changes, your dashboard will now appear.

Summary

Dashboards can provide great visuals to allow users to quickly see information across entities. This chapter outlined the basic information you will need to create your own charts and add them to new or existing dashboards. Utilizing dashboards and giving access to specific user groups to information they will need will greatly improve the value of the data within your Dynamics 365 environment.

Chapter 11
Plugins and Workflow Activities

With all of the customizations and out-of-the-box features available with Dynamics 365, you will still encounter challenges that will require you to write code in C#. C# code that can be called from events triggered within CRM entities, such as inserts, updates, and deletes, are contained within Plugins. This chapter will outline how to develop, deploy, and administer plugins and workflow activities.

You may not be a developer, but it is important to understand the basic approach to developing these artifacts. Make sure you understand what is involved in a plugin or custom workflow activity, so if you decide that you need one for your environment, you know what you are asking for and how these components operate and relate to Dynamics.

Developing a Plugin

Before you do development, you should download the Microsoft Dynamics 365 Software Development Kit (CRM SDK). This SDK contains a number of assemblies that you will need to reference in your plugin and workflow activity development. You can download the SDK from the URL below (also shown in Figure 11.1). While you can do your plugin development without downloading this directly (you can reference NuGet packages from within Visual Studio, as will be shown later), you will need access to the tools in this SDK in order to deploy your plugin.

```
https://www.microsoft.com/en-
us/download/details.aspx?id=50032
```

DOI 10.1515/9781547400065-011

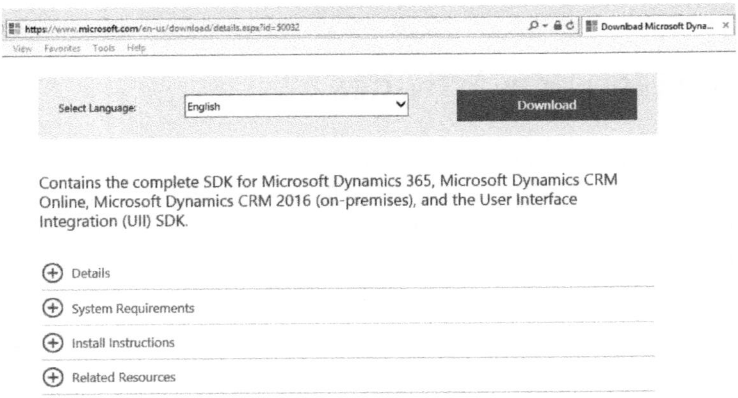

Figure 11.1: Download the SDK

To create a Plugin, open Visual Studio 2017. Create a new C# Class Library, as shown in Figure 11.2. You will want to rename it from the default ClassLibrary1 to something more descriptive. Click OK and Visual Studio will create a new project and solution for you.

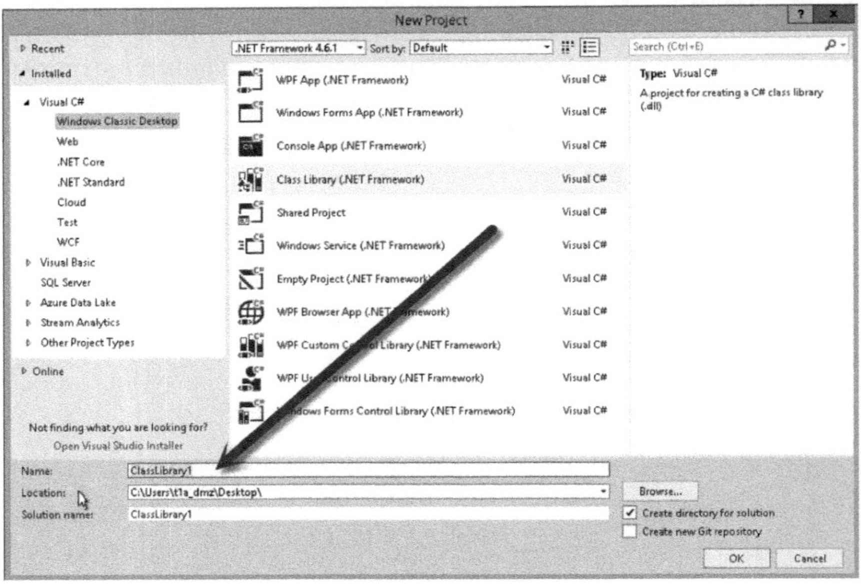

Figure 11.2: Creating a C# Class library for the plugin project

Now click on the Tools menu and select the NuGet Package Manager and then Manage NuGet Packages for a Solution (see Figure 11.3). This will allow you to reference the latest CRM SDK assemblies (you can also browse to individual assemblies on your local computer if you have downloaded the full CRM SDK described above).

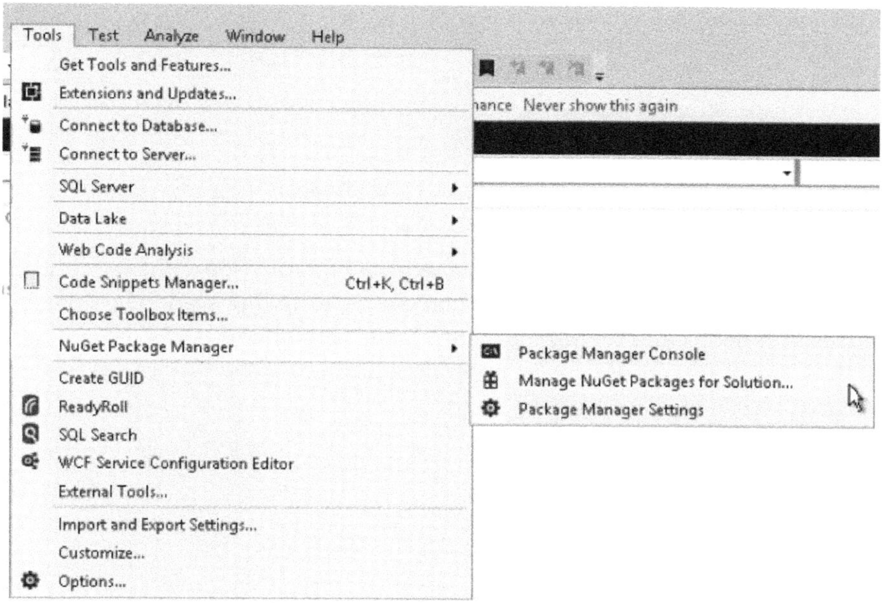

Figure 11.3: There are NuGet Packages available for the SDK

When browsing for the NuGet package to reference, type "CRM core" in the search interface. You are looking for the Microsoft.CrmSdk.CoreAssemblies package (Figure 11.4). This can be found by typing the search text in the box under the Browse header.

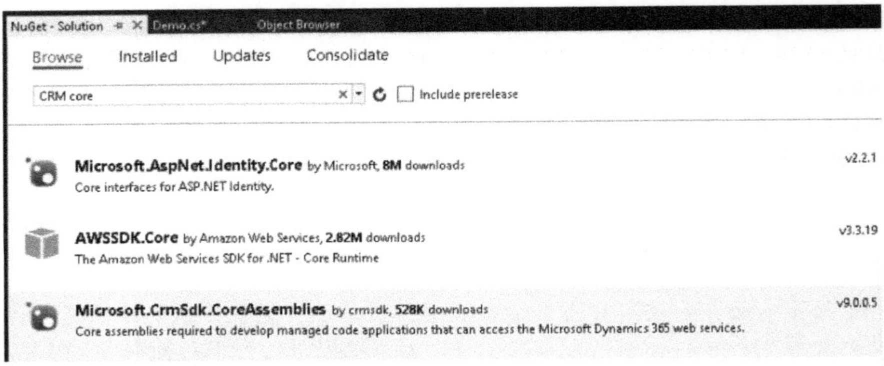

Figure 11.4: The CoreAssemblies package has what you will need for plugin development

Once you find the right package, click the Install button. Notice the description (Figure 11.5) on this package. It shows that you are installing everything from the Microsoft.Xrm.Sdk and Microsoft.Crm.Sdk.Proxy assemblies.

Figure 11.5: Version and assemblies included

When the package has downloaded and installed, all the assembly references you need for building your plugin will have been added to your project. Figure 11.6 shows the full project, along with all files and references.

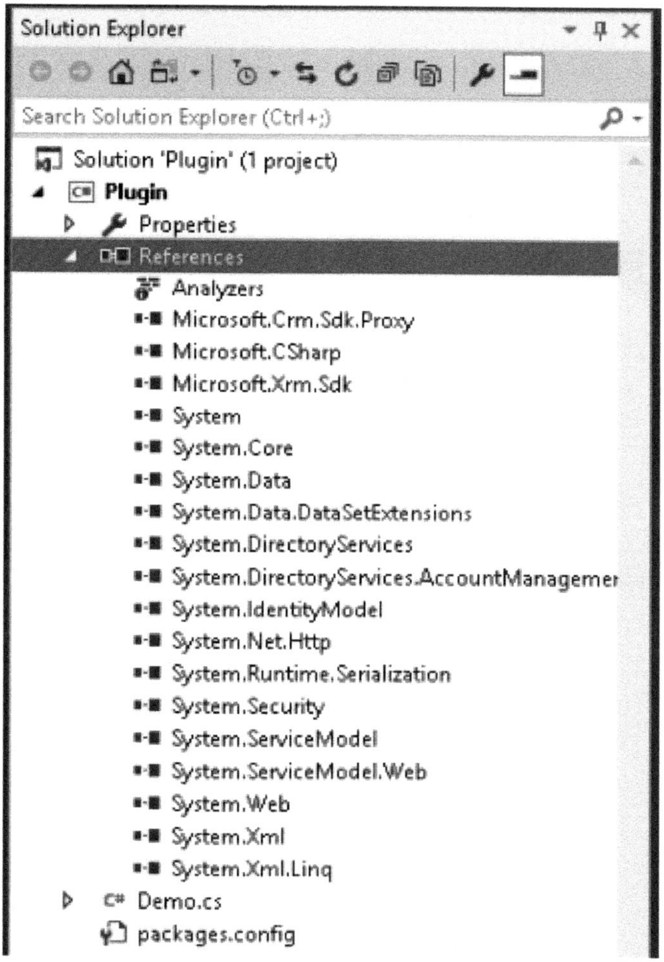

Figure 11.6: The full project with files and references to the SDK assemblies

Now you can begin coding. Open the .cs file in your project and enter the code shown in Listing 11.1. This listing shows the bare minimum code that must be in place in order to create a functional plugin. This code has no actual work being done in it, but it does contain the framework. Your plugin-specific logic would go inside of the Execute method.

Listing 11.1. Framework C# for a Functional Plugin

```
using System;
using System.Collections.Generic;
using System.Linq;
using System.Text;
using System.Threading.Tasks;
using Microsoft.Xrm.Sdk;
using Microsoft.Crm.Sdk;
namespace Plugin
{
 public class Demo : IPlugin
 {
 public void Execute(IServiceProvider service)
 {
  // plugin logic goes here, but plugin will build and can be
deployed with
 // only the code shown here.
 }
 }
}
```

A plugin's base class must inherit from the IPlugin.interface, as shown in Figure 11.7.

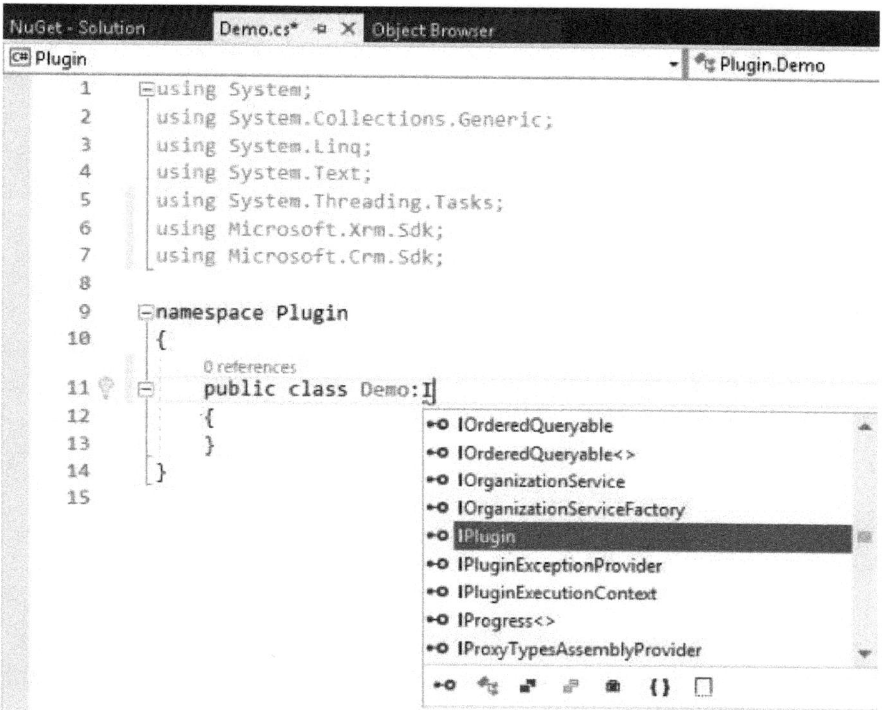

Figure 11.7: Plugin base class inherits from IPlugin

With the code added to the .cs file, you have one remaining thing to do prior to building and deploying your plugin. You must add a strong name key to the solution. You can do this by right-clicking your project in the Solution Explorer tab and selecting Properties. On the Properties tab, click on Signing and Create a new strong name key. Figure 11.8 shows this being done.

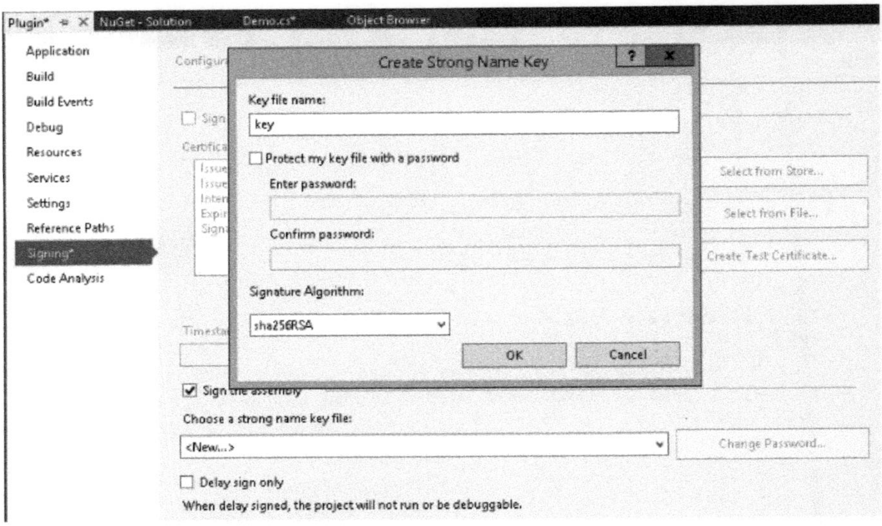

Figure 11.8: Adding a strong name key to your project

Finally, you can right-click your solution and build it. This will create an assembly (see Figure 11.9). The assembly can now be deployed to Dynamics 365, as outlined later in this chapter.

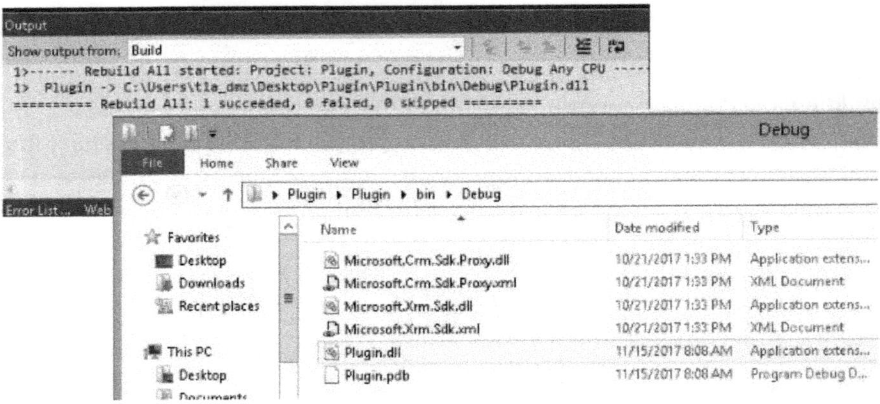

Figure 11.9: The final plugin assembly

Developing a Workflow Activity

A very basic overview of workflow activities and how they can be referenced from workflows was outlined in Chapter 7. This section will go into detail on how to develop a fully functional workflow activity. The workflow activity will be used to pull a configurable value from a web resource file. This can be invaluable in scenarios where you have a value that must be set in a large number of workflows, and that may change as you move these workflows between environments. For this example, let's say you have a development environment and a production environment. You have a workflow that references an SSRS report server that is used in the logic of one of the steps in your workflow. You need to be able to configure this workflow to point to the development SSRS server in one environment and the production SSRS server in the other. To configure this value, you have to type a hard coded value into the property of a step in the report. You have over 50 different workflows where this must be configured, so every time you do a deployment from development to production, you have to manually alter each of these workflows. What you want is a way to change this configuration in one location, through the use of a global variable accessible to the workflows.

The only way to have a configurable field that pulls from an external source in a workflow is to call a workflow activity. This workflow activity can then pull the information from a web resource. In this example, we'll start by creating a web resource where the global variable can be stored by taking the following steps:

1. Click on Web Resources in your solution and click on the New button, as shown in Figure 11.10.

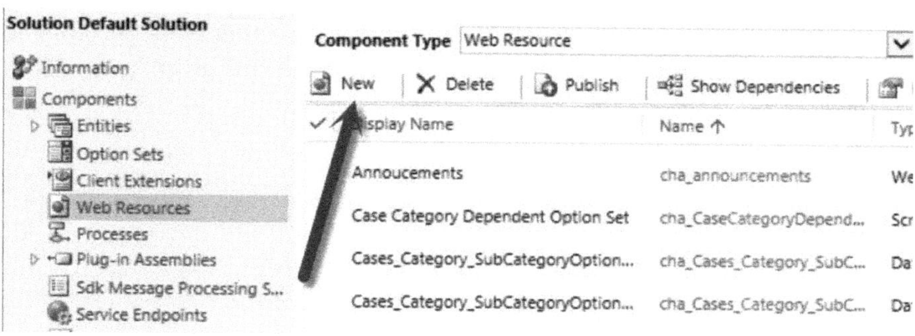

Figure 11.10: Creating a new web resource

2. In the window that opens, give your web resource a name, then select the Type as a Script (Jscript), and click the Text Editor button (see Figure 11.11).

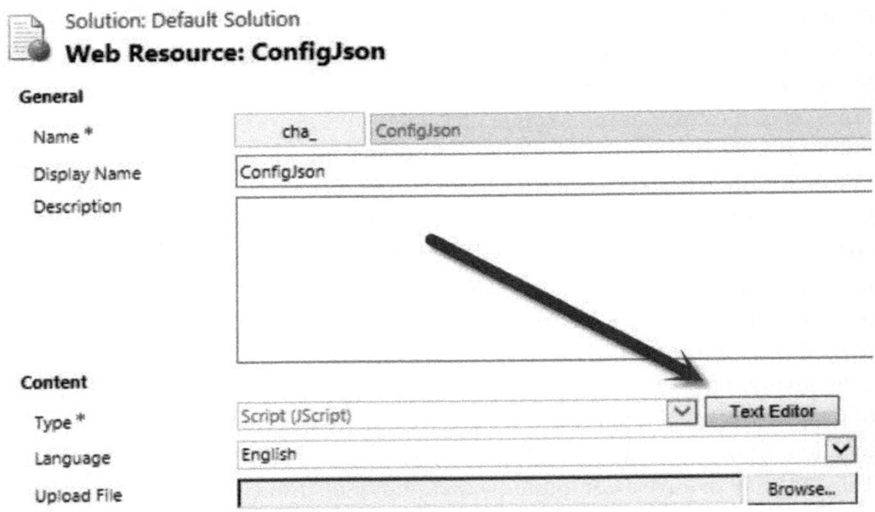

Figure 11.11: Click the Text Editor to enter in the script

3. In the screen that opens, you can type the information for your global variable. There are a number of formats you could use to do this, but in this example the variable is stored as shown in Figure 11.12. This string will be read by the C# in the workflow activity. The variable we are after is the second one shown, called ReportServer. The "http" address will change in the two environments.

Figure 11.12: Two parameters are configured

Once you have entered the code, you can save and publish your new web resource. The next step is to create your workflow activity. You need to create a new class library C# project in Visual Studio, add a strong name key, and then code the C#. The solution in Visual Studio will look similar to that shown in Figure 11.13.

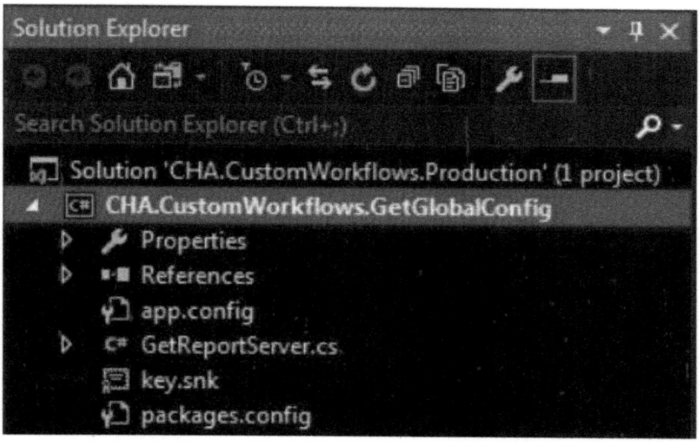

Figure 11.13: The solution in Visual Studio

The code for the workflow activity in its entirety is shown in Listing 11.2. There are several things of importance in this code—review the comments in this listing to understand what is taking place. The function of this code is to retrieve the value from the web resource just created and return it to the workflow that is calling it.

The primary difference between a plugin and a workflow activity is that the base class for a plugin inherits from the IPlugin interface, whereas the base class for a workflow activity inherits from CodeActivity. This is shown in Listing 11.2 as public class GetReportServer : CodeActivity.

Listing 11.2. Workflow Activity to Read Value From Web Resource

```
// there are a number of assemblies that must be referenced
by your solution
// all of these are available in the CRM SDK that you
downloaded
// you can find them in the BIN folder of the extracted SDK
files
using System;
using System.Collections.Generic;
using System.Linq;
using System.Text;
using System.Threading.Tasks;
using System.IO;
using System.Runtime.Serialization.Json;
using Microsoft.Crm.Sdk.Messages;
using Microsoft.Xrm.Sdk;
using Microsoft.Xrm.Sdk.Messages;
using Microsoft.Xrm.Sdk.Query;
using Microsoft.Xrm.Sdk.Workflow;
using System.Activities;
// Make sure and give your assemblies intelligent namespaces
namespace CHA.CustomWorkflows.GetGlobalConfig
{
 // inherit from CodeActivity
 public class GetReportServer : CodeActivity
 {
 // this is your output parameter
 [Output("ReportServer")]
 public OutArgument<string> ReportServer { get; set; }
 protected override void Execute(CodeActivityContext
executionContext)
 {
 ITracingService trace =
executionContext.GetExtension<ITracingService>();
 IWorkflowContext workflowContext = executionCon
 text.GetExtension<IWorkflowContext>();
 IOrganizationServiceFactory serviceFactory = executionCon
 text.GetExtension<IOrganizationServiceFactory>();
 IOrganizationService service = serviceFacto
```

```
ry.CreateOrganizationService(workflowContext.UserId);

// here the code tries to access the web resource that you
have created
try
{
Entity WebRef = null;
WebRef = RetrieveEntity(service, "webresource", "name",
"cha_ConfigJson");
if (WebRef.Attributes.Count > 0 )
{
string webresource = getStringContent(WebRef);
Config config = ConvertJSONToObject(webresource);
ReportServer.Set(executionContext, config.ReportServer);
}

}
catch (Exception ex)
{
trace.Trace("CHA.CustomWorkflows.GetGlobalConfig.GetBAseURL
Failed: Error Mes
sage: " + ex.ToString());
throw new InvalidPluginExecutionException("Error Message: "
+ ex.ToString());
}
}

// this will retrieve the web resource
public Entity RetrieveEntity(IOrganizationService service,
string EntityName,
string field, string value)
{
Entity e = new Entity();
Guid id = new Guid();
QueryExpression query = new QueryExpression();
query.EntityName = EntityName;
query.ColumnSet = new ColumnSet(true);

FilterExpression fi = new FilterExpression();
fi.AddCondition(new ConditionExpression(field,
ConditionOperator.Equal, value));
query.Criteria = fi;
RetrieveMultipleRequest retreiveRequest = new
RetrieveMultipleRequest();
retreiveRequest.Query = query;
RetrieveMultipleResponse response = (RetrieveMultipleRe
sponse)service.Execute(retreiveRequest);
if (response.EntityCollection.Entities.Count > 0)
{
e = response.EntityCollection.Entities[0];
```

```
}
return e;
}


public string getStringContent(Entity WebRef)
{
string content = string.Empty;
if (WebRef.Contains("content"))
{
byte[] b =
Convert.FromBase64String(WebRef.Attributes["content"].ToStri
ng());
content = UnicodeEncoding.UTF8.GetString(b);
}
return content;
}

public Config ConvertJSONToObject(string webresource)
{
Config config = new Config();
string JSON =
webresource.Substring(webresource.IndexOf('{'),
(webresource.Length
- webresource.IndexOf('{')) - 1).Replace("\\", "");
MemoryStream ms = new
MemoryStream(Encoding.UTF8.GetBytes(JSON));
DataContractJsonSerializer ser = new
DataContractJsonSerializ
er(config.GetType());
config = ser.ReadObject(ms) as Config;
return config;
}
}

public class Config
{
public string IFrameRoot;
public string ReportServer;
}
}
```

Once you have your C# written, you need to build your project in Visual Studio and make sure that an assembly is created. You will now be able to deploy this assembly to CRM and reference it from a workflow activity. See the next section in this chapter to work through the deployment process. When the assembly has been deployed, you will be able to reference your new assembly from the Step list in a workflow (see Figure 11.14).

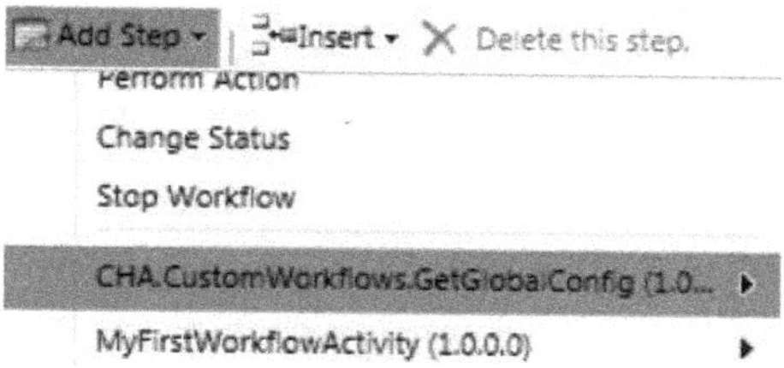

Figure 11.14: The workflow activity is available as a step once it has been deployed

Adding this workflow activity step will create a single step in your workflow that looks like the one shown in Figure 11.15. You will generally want to add this type of lookup to a variable as the first step in your workflow to ensure you have access to the value in other steps in your workflow.

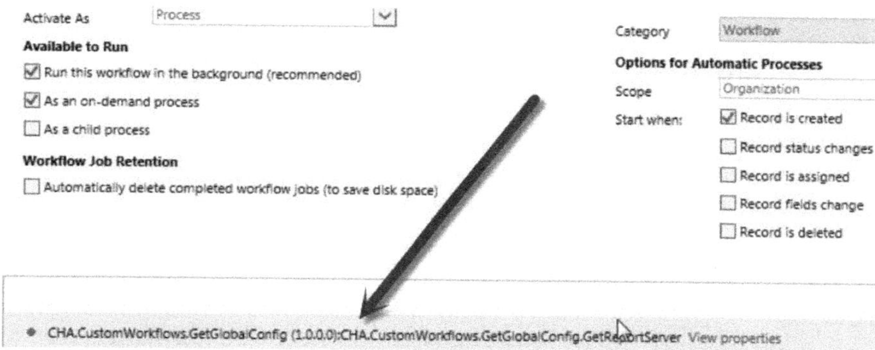

Figure 11.15: The workflow activity is set as the first step in the workflow

This workflow activity's returned value can now be accessed directly in other step's configurations by accessing it from the form assistant. Figure 11.16 shows this value being added to the Report Server property of a separate step in the same workflow.

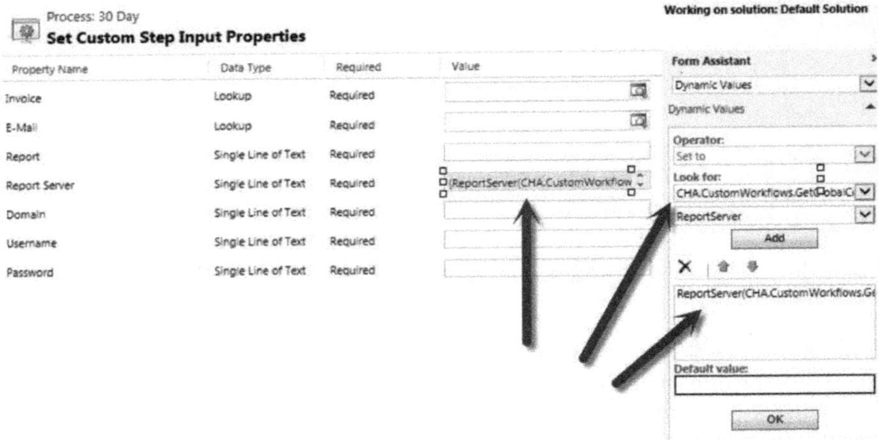

Figure 11.16: Adding the dynamic value to a property within a step of the workflow

Deploying Plugins and Workflow Activities

The deployment process for workflow activities and plugins is very similar. In order to deploy these assemblies, you must use the Plugin Registration Tool, which is included in the CRM SDK. This tool can be found in the extracted files in the SDK, as shown in Figure 11.17.

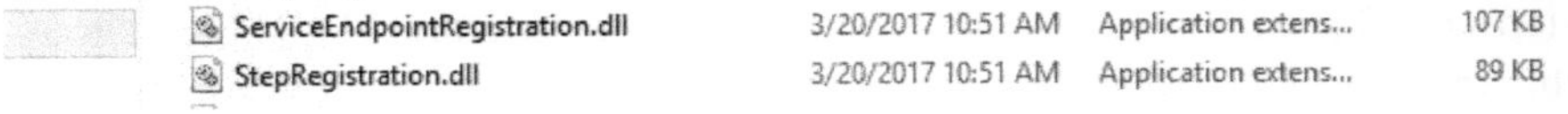

Figure 11.17: Accessing the Plugin Registration tool, included in the SDK

When you run this tool and it opens, you will need to begin by clicking the Create New Connection button. This button will open a box where you must select Office 365 as the deployment type and then enter your D365 credentials, as shown in

Figure 11.18. You need to select the Show Advanced button in order to be able to enter your information. Enter your information, and then click the Login button.

Creating a connection may not work as easily as you expect. The Registration Tool will search to see whether you have any active browser sessions with a connection to Dynamics 365 already in place. If you do, that session will override what you may type into the tool's connection information. If you get an error saying your credentials are invalid, but you know you typed the correct thing, try closing all your browser sessions.

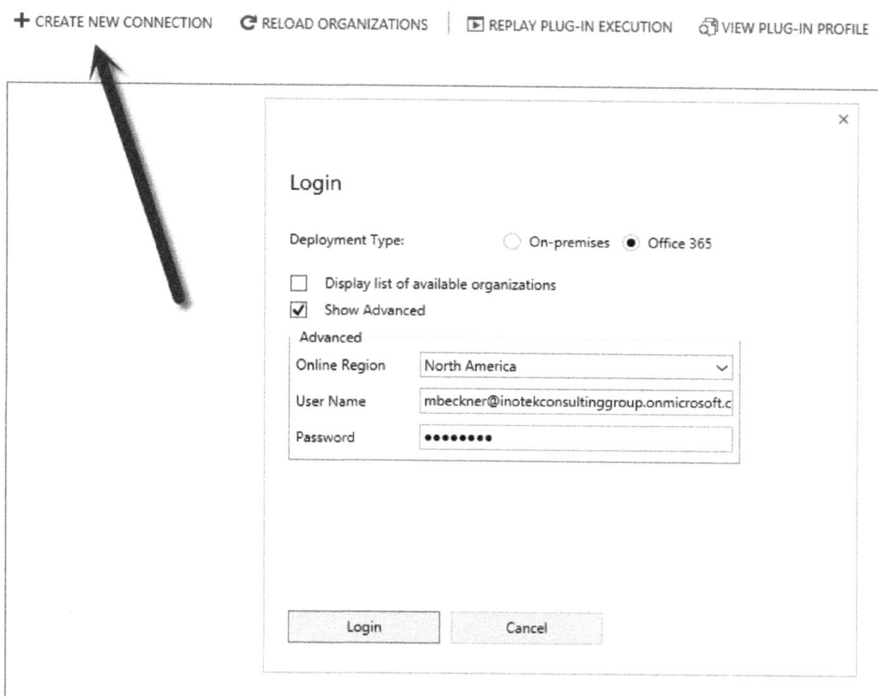

Figure 11.18: Creating a connection to Dynamics within the Plugin Registration tool

Once you are logged in, the screen will show a list of all of the plugins and workflow activities that have been deployed. You can expand each of these to see details about them. In the case of plugins, you will have a few extra items to configure over a workflow activity. You can deploy both of them by clicking the Register button and then selecting Register New Assembly, as shown in Figure 11.19.

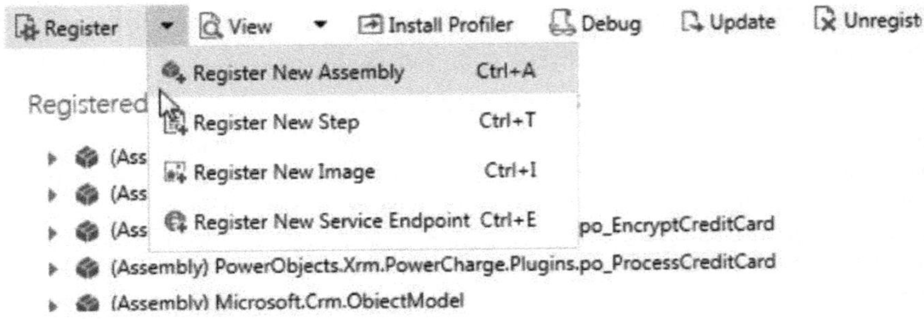

Figure 11.19: Register a new assembly

A screen will open where you can configure the details around the deployment. The plugin registration tool will take the assembly and place it into the Dynamics environment. Set the properties shown in Figure 11.20 as follows:

1. In Step 1, browse to the location of your compiled assembly, choose the assembly, and click the Load Assembly button.
2. In Step 2, select the plugins or activities you are planning to register. Some developers choose to have a single assembly with many plugins in it. Others choose to have one plugin or workflow activity per assembly. In the case shown here, there is just one workflow activity in the assembly, so it is automatically selected.
3. In Step 3, you should select the Sandbox option. When you deploy assemblies, you will generally deploy to the Sandbox. In some special cases, especially with On Premise deployments, you may want to have no isolation. For example, in an On Premise environment, you may want to interact directly with a SQL database. You cannot do this in Sandbox, so you would select None. In an Online environment, you will almost always use Sandbox.
4. In Step 4, select Database. The other options are for On Premise deployments.

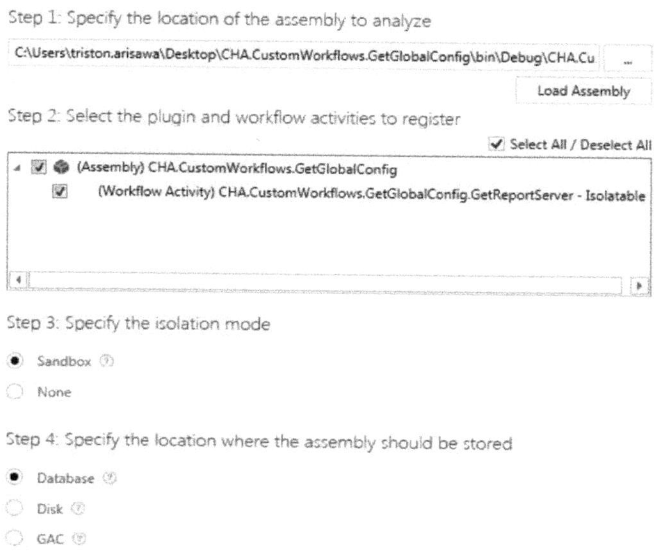

Figure 11.20: Configuring the assembly deployment

When you click OK on this screen, you will have completed the deployment, and your assembly will now appear in the list of assemblies. If it is a workflow activity assembly, your deployment is complete. If it is a plugin assembly, there will be several additional steps you need to take. These steps include registering a new step and registering an image. For a plugin that has no properties being passed into it, you can simply register a new step. A new step allows you to configure when this plugin will fire. If, for example, you wanted to configure your plugin to execute on the Create event of an account, you would configure your step similar to the one shown in Figure 11.21. Details on the key fields are:

- Message—This is the event that you want to trigger on. The most common are Create and Update, though there are many other options that will appear as you type in this box. In this case, we are listening for the Create event.
- Primary Entity—This is the entity that you want the plugin to be attached to. In this case, it will be the Account entity.
- Run in User's Context—If you want this plugin to fire the same for everyone, you'll want to set this value to a specific user that has System Administrator privileges. If you want it to fire based on the permission of the current user, set it to Calling User. For example, let's say your plugin was going to create a new entity based on the data in the newly created account. You may have a

user that does not have permission to create this secondary entity. If the plugin fires under the "User Context," then it would cause an error saying it didn't have appropriate permission. In order to force the plugin to run under the same administrative account for everyone, you would want to set it to the administrator account so that the plugin could fire and execute the code the same, regardless of the current user.

- Event Pipeline Stage of Execution—You can have the plugin fire before the event or after the event. In this case, we want the plugin to fire after the event, so that the account has already been created by the time the logic in the plugin fires.
- Execution Mode—Asynchronous execution allows the plugin to fire while the user moves on with their work. Synchronous means the plugin must complete before the user is allowed to move on with using Dynamics.

When you are done configuring the step, click the Register New Step button. This will add the step under the assembly. You can configure additional steps (perhaps for the Update or Delete Event on the same or different entity).

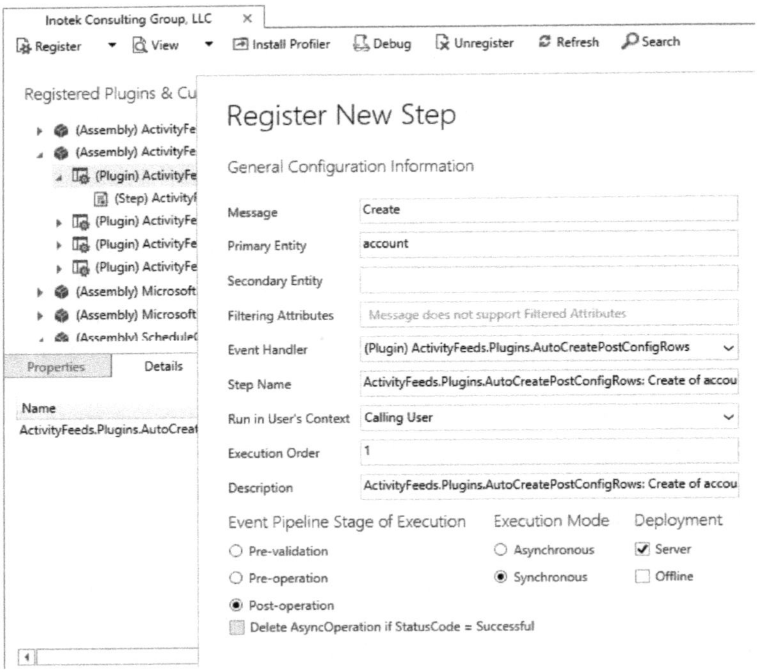

Figure 11.21: Registering a step in a plugin

Administering Assemblies

Plugins and Workflow Activities are both assemblies. These assemblies, once deployed to Dynamics, can be viewed directly from the Dynamics web interface. If you want to see what plugins and workflow activities have been deployed, click on the Plug-in Assemblies under the Components header in your solution, as shown in Figure 11.22.

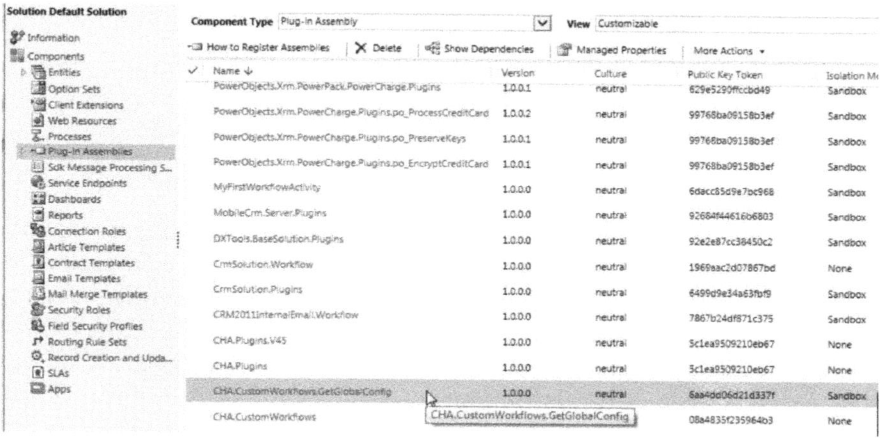

Figure 11.22: Accessing plugins from within a solution in Dynamics

You can delete assemblies from this view, but you can't do much else of consequence. For most of the administration related to these assemblies, you will continue to use the Plugin Registration Tool. You can, however, enable or disable the plugin steps. To do this, click on the Sdk Message Processing Steps tab, right under the Plug-in Assemblies tab (see Figure 11.23). Select the plugin with the step that you are interested in and then click the Activate or Deactivate button on the toolbar.

Disabling plugins may be required during testing or during large imports of data. For example, you may want to import 10,000 new contacts from a spreadsheet. There may be a plugin that fires on creation of a contact record that you do not want to execute when these external contacts are loaded. You can easily disable a plugin by deactivating the Create step in the Sdk Message Processing Steps.

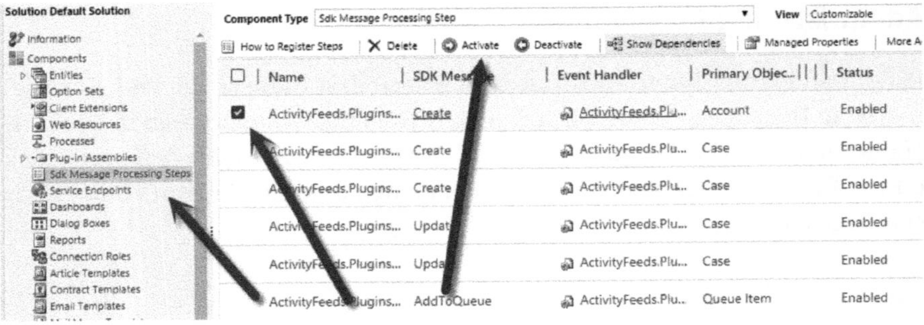

Figure 11.23: You can activate or deactivate individual steps within plugins

Summary

This chapter was much more developer-centric than previous chapters in this book. It is important for anyone working within a Dynamics 365 environment to understand the tools and processes involved in developing, deploying, and administering plugins and workflow activities, even if not a developer. Often, business analysts and project managers can't speak directly to what a coder is doing, and treat things written in code like black boxes. To be able to really architect and direct a system, it is important to understand the basic concepts and pieces to the platform, including those that are primarily code-based.

Chapter 12
Ribbons

Throughout the various entities within CRM, you have access to a variety of buttons and functionality available on the toolbar or menu "ribbon." These ribbons are unique to each entity and can be customized. Customization does require code and a special toolkit that you will be introduced to in this chapter. Like the previous chapter on plugins, this chapter is more developer focused, but once again it is important that you understand what can be done within the platform, and have a basic idea of the complexity associated with the task of modifying these ribbons.

In this chapter you will learn how to add a button to the account entity's ribbon. The result of this will look like Figure 12.1.

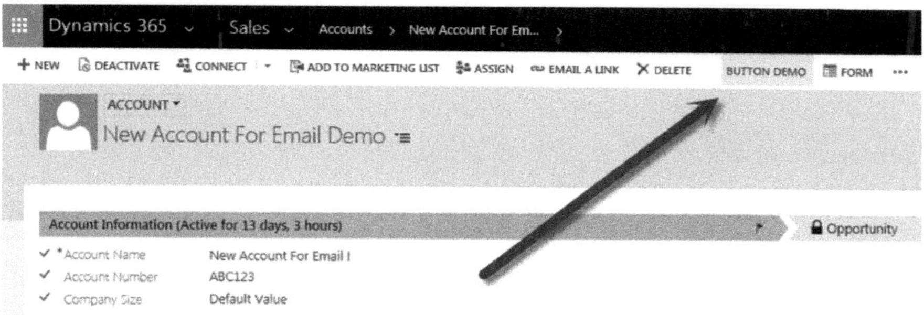

Figure 12.1: You will work through creating a button on the account entity.

Start by creating a new solution, and then adding the account entity to this solution. There should be a single entity in the solution, as shown in Figure 12.2. See Chapter 3 for details on creating an entity and Chapter 8 on creating a new Solution.

DOI 10.1515/9781547400065-012

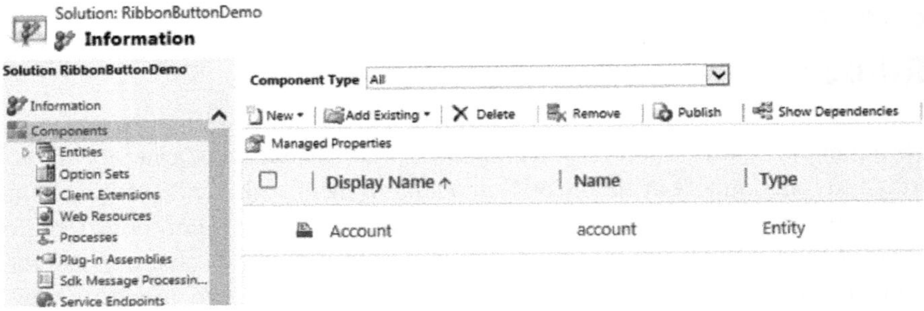

Figure 12.2: Create a new solution—this will be used by the ribbon workbench shortly.

Next, create a new web resource file, which is of type Script (Jscript). Click New, then Web Resource. After entering the Name and Display Name, select Script (Jscript) from the Type drop-down list. Then click the Text Editor button and add the simple alert code that is shown in Figure 12.3. After you have entered this code, save and publish this new solution.

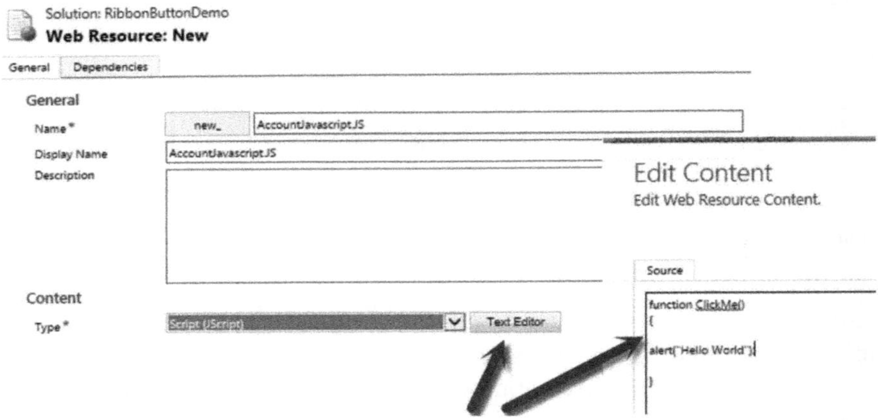

Figure 12.3: A simple JavaScript method that will be called from the button when clicked.

Using the Ribbon Workbench

While you can modify a ribbon using XML and importing the XML file into the solution, the easiest and most effective way to do ribbon development is with the

Ribbon Workbench. This can be downloaded at the following URL, and the download page can be seen in Figure 12.4.

```
http://www.develop1.net/public/rwb/ribbonworkbench.aspx
```

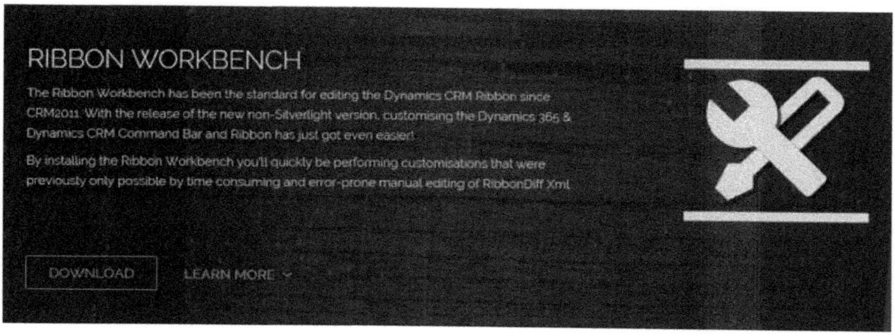

Figure 12.4: Downloading the Ribbon Workbench.

When you download the workbench, you will get a .zip file. This .zip file must be imported into CRM, just like you would do with any managed solution. To do this, click on the Import button in your solutions, as shown in Figure 12.5.

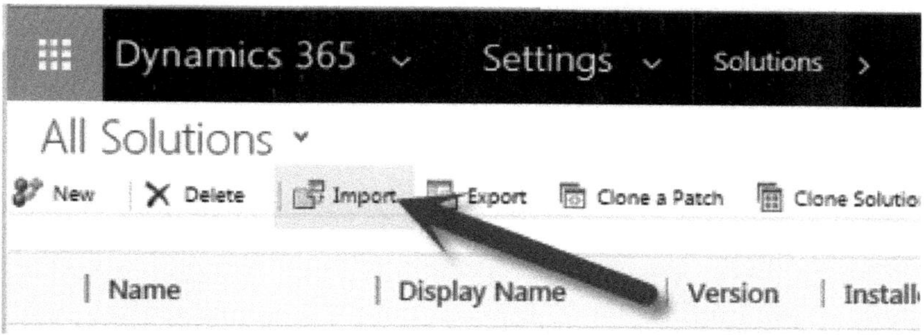

Figure 12.5: Importing the workbench into Dynamics.

Clicking the Import button will open the import wizard. You should be able to navigate through this wizard to the downloaded .zip file. Select the .zip file, and click Next at the bottom of the dialog box, then Next again, then Import. ==This will cause the Ribbon Workbench to be imported into your solutions list (see Figure 12.6). You may have to refresh your browser to see the button.

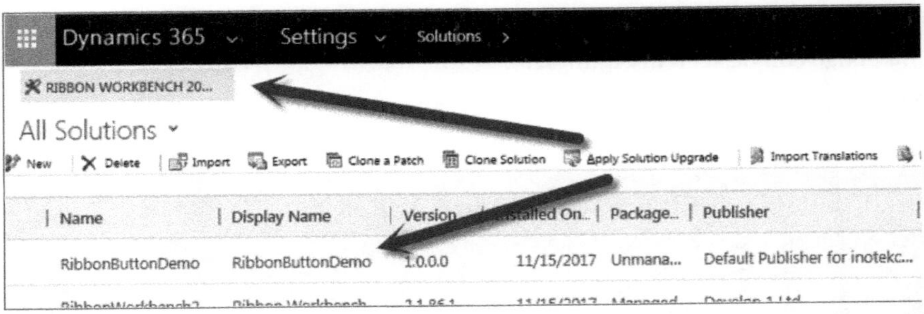

Figure 12.6: The import will cause a new button to appear after browser refresh.

Click the Ribbon Workbench button at the top of the solution window, and in the Ribbon Workbench window that opens, click Open Solution (see Figure 12.7). Browse to the solution you created earlier in this chapter and click OK. This will allow you to make modifications to your account entity.

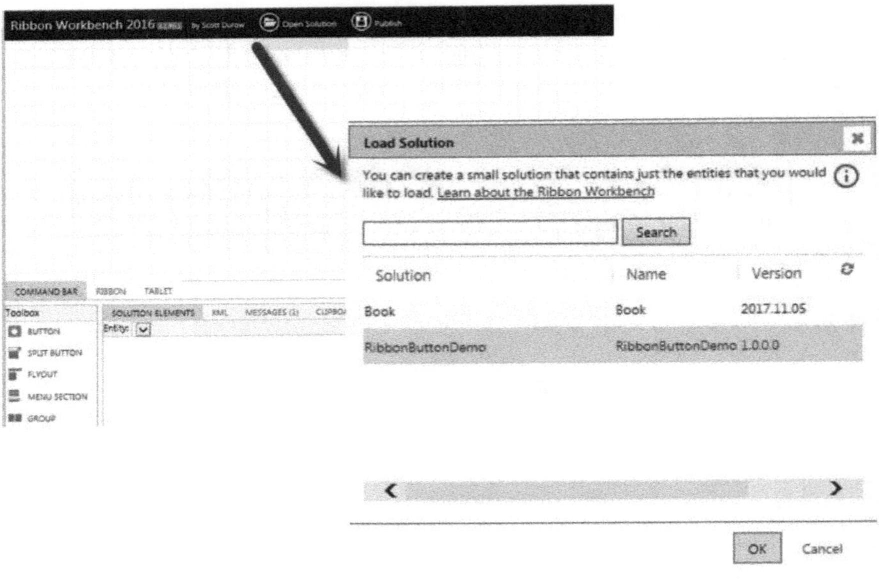

Figure 12.7: Use the workbench to Open the solution you just created.

You will now see a fairly complex editor. Your existing ribbon will be shown on the upper section, while the toolbox and various options with what you can do to the ribbon will be in the lower section. For this demo, you will add a button to the right of the existing "Email A Link" button. To do this, drag your button from the Solution Elements under the Ribbon tab on the bottom left and drop the button to where you want it on the toolbar. When you have dropped the button, name it in the Properties: Control section that appears in the lower right of the screen. All of this can be seen in Figure 12.8.

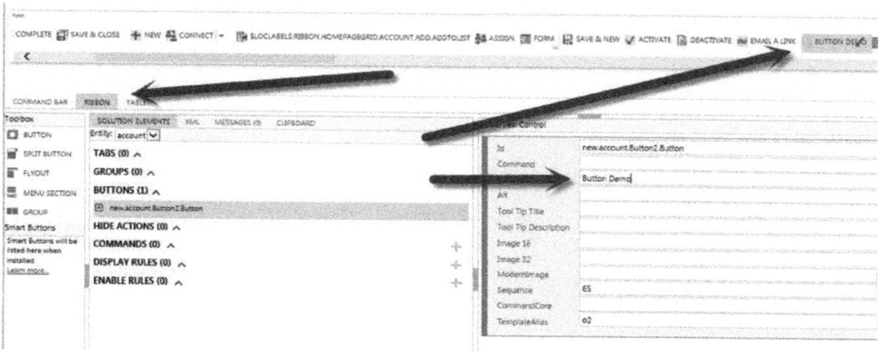

Figure 12.8: Adding the button to the ribbon and naming it.

When you have the button renamed, expand the COMMANDS header in the Solution Elements tab and then click on Add Action to the right. Clicking the Add Action button will allow you to then select JavaScript Action from a small menu (refer to Figure 12.9).

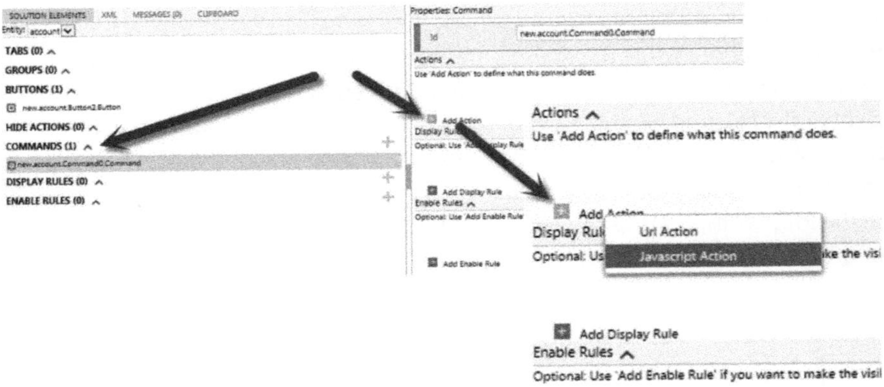

Figure 12.9: Adding a JavaScript action to the button.

In the Library property, type in the first few letters of your web resource name (or use the lookup tool to locate it). Figure 12.10 shows the new_AccountJavascript.JS, which is the JavaScript that you published earlier in this chapter.

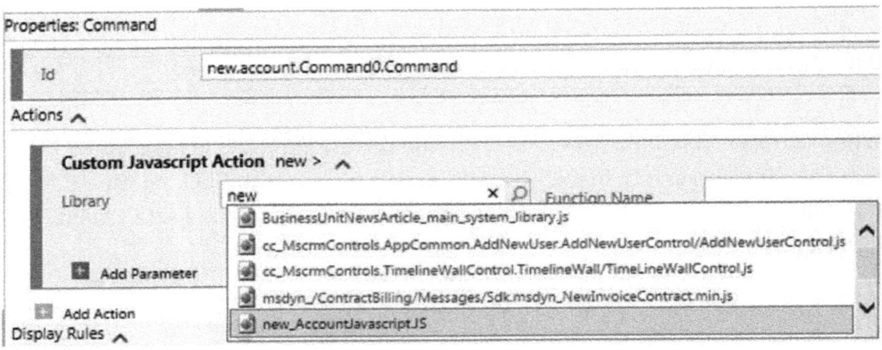

Figure 12.10: Referencing the JavaScript web resource.

You will do the same for the Function which should be "ClickMe." Figure 12.11 shows this JavaScript box fully configured.

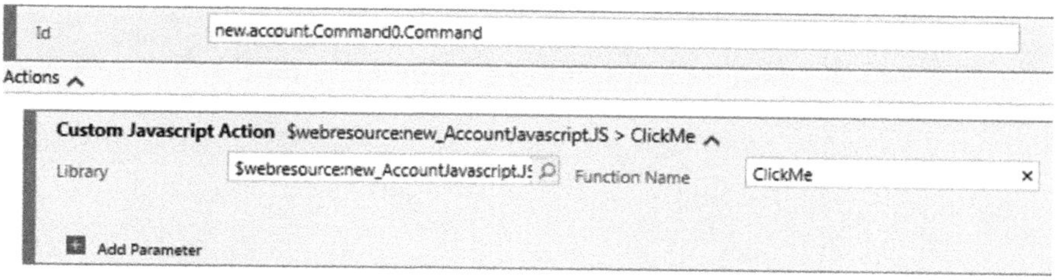

Figure 12.11: The library and function configured.

Back in the properties of the button, you will now need to click the Command property's drop down and select the command. This will be the command you just created with the reference to the ClickMe function. Figure 12.12 shows this property being set.

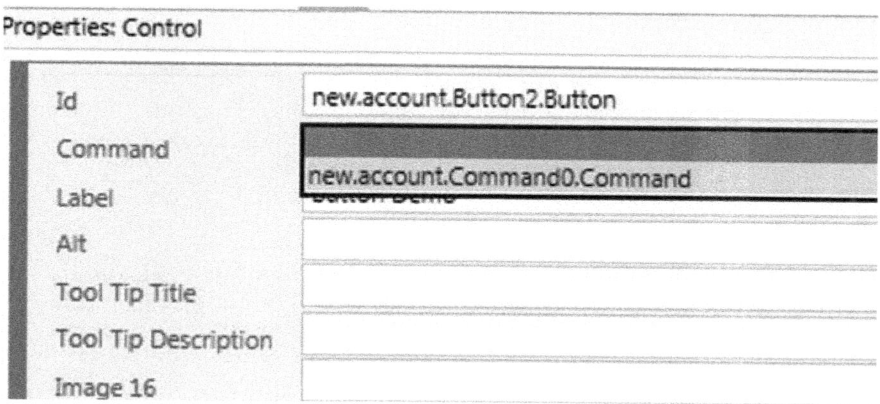

Figure 12.12: Adding the command reference.

You have now added a button that calls JavaScript. You can now click on the Publish button in the Ribbon Workbench (see Figure 12.13). This will take a bit of time to complete.

Figure 12.13: Publishing your changes from within the workbench.

Close out of the Ribbon Workbench and return to your main Dynamics 365 browser window. Refresh this window, and open Accounts. You should see the new button on your account form. When you click the button, you should see the alert pop up (see Figure 12.14).

Figure 12.14: Clicking the published button.

Summary

The Ribbon Workbench allows you to do just about anything you need to with an entity's ribbon. This chapter demonstrated a very simple button that calls JavaScript, but your button can do whatever is needed, including opening a dialog, initiating a workflow, opening a report, exporting records, and more. Knowing the basics to ribbon development will allow you to incorporate custom functionality on your entities with relative ease.

Index